# Precious Water
## in Rocky Mountain National Park

**Water, Ditches, Dams & Floods**
**The 1982 Lawn Lake Flood**

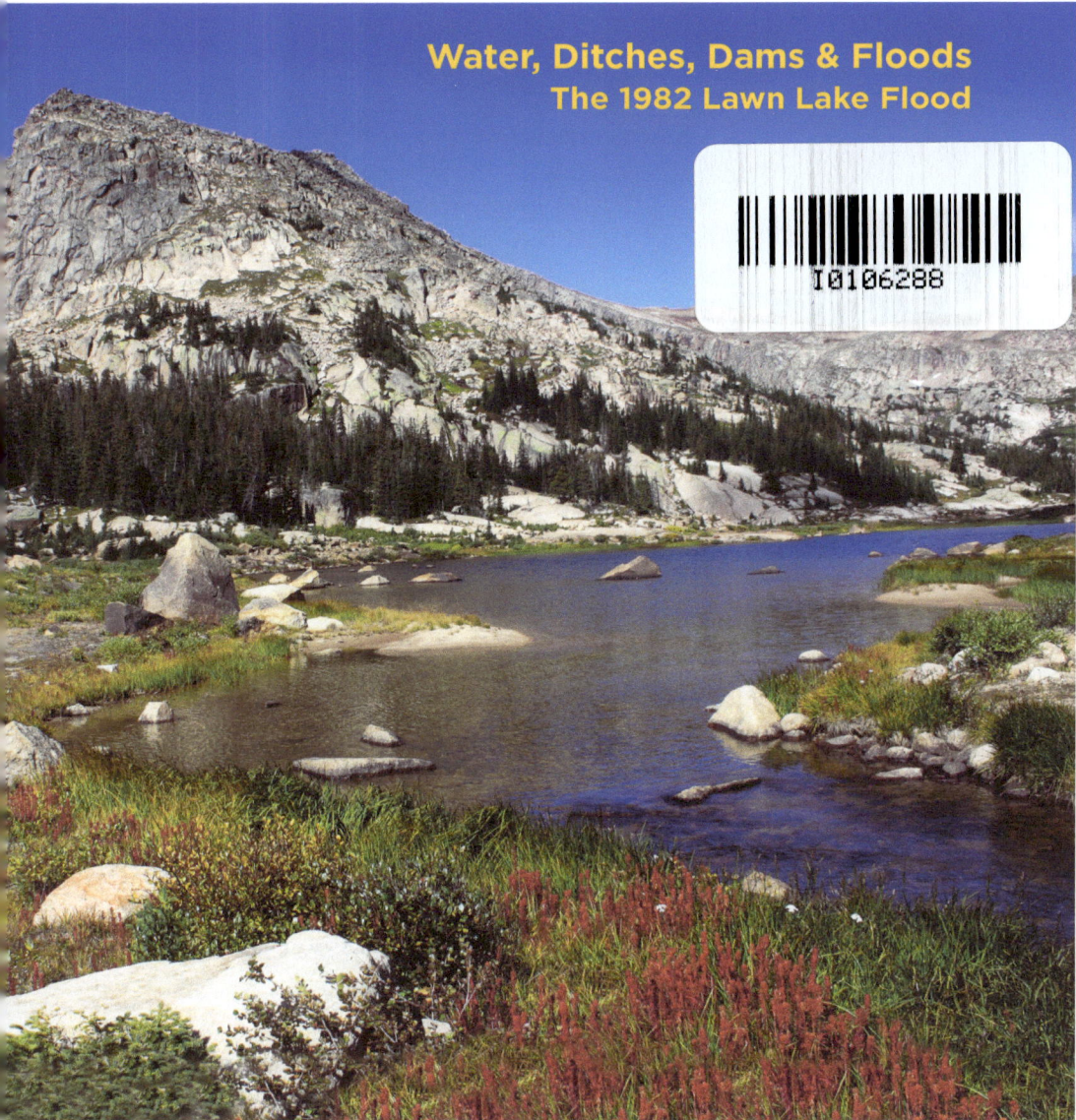

**Daniel N. Gossett**

**Precious Water In Rocky Mountain National Park**
*Water, Ditches, Dams and Floods*
*The 1982 Lawn Lake Flood*
© 2022 Daniel N. Gossett

As a writer of this history and these events, I have relied on documents written
by many others in the past 40 years and beyond, to complete this book. In my
attempts to put all this together, I may have invariably missed some credit to
people where it is due and made mistakes in my details. These mistakes and
omissions are my own. I apologize to those where I have erred.

All photographs including the cover images of Lawn Lake and
Brook Trout in Boulder Brook are by the author unless otherwise noted.

Cover and Interior Design: Rebecca Finkel, F + P Graphic Design, FPGD.com
eBook conversion: Rebecca Finkel, F + P Graphic Design, FPGD.com

Library of Congress Control Number: 2022908672
ISBN IS hardcover: 979-8-9861586-0-0
ISBN IS paperback: 979-8-9861586-3-1
ISBN KDP paperback: 979-8-9861586-1-7
ISBN eBook: 979-8-9861586-2-4

Rocky Mountain National Park  |  Water Resources  |  Lawn Lake Flood

First Edition
Printed in the USA

*This book is dedicated to those lost to the floodwaters,*

*Steven See*

*Bridget Dorris*

*Terry Coates*

*and to Steven Gillette,*

*whose fast action saved countless lives,*

*and to the National Park Service*

*employees and first responders*

*who mobilized following the Lawn Lake Flood.*

"Whiskey is for drinking, and water is for fighting."

—ATTRIBUTED TO MARK TWAIN

# Contents

# Introduction

During my first summer with the National Park Service (NPS) in 1974, our Trails Boss, Jack Gartner and I had finished unloading the pack horses for a ten-day campout for my trail crew at Finch Lake. He and I then hiked on up to Pear Reservoir, a beautiful alpine lake 6.2 miles in Wild Basin on the southern edge of Rocky Mountain National Park (RMNP). Jack was standing with me on the top of the deteriorating 1906 dam structure, and he told me of his idea of blowing the dam to remove it. The dam was one of four Arbuckle Reservoirs built in the early 1900's that became water storage for the City of Longmont, and most of them were in disrepair, with leaks and holes in them pouring out into the creeks in lieu of functioning outlets or spillways. Pear Lake was supposed to be back to its original pear shape, not filling to the dam, except for flowing out into Cony Creek.[1] I recall that water was up to the dam when I was there, I believe due to early season snowmelt filling the lake.

Jack told me that blasting three feet of dam at a time would make small enough rubble that dispersal via explosion would be unnoticeable in the high valley. Jack was one of two NPS employees who handled explosives, the other was the Road Crew supervisor, Ken Hockleberg. Later this duty would go to George Wagner, the East District Office Ranger, who took over Jack's responsibilities overseeing trail crews. The Park Service dynamite was securely stored in a lockbox near the Park Headquarters' Utility Area.

My last summer as a seasonal trail crew worker was in 1980. I had just finished paramedic school and started back at the Park late in the season. I was working with different trail crews in areas besides the southern trails of Wild Basin and Long's Peak, where most of my RMNP seasonal career had been spent. I was helping Brady Wheeler's trail crew drilling rocks for blasting on the Pott's Puddle trail, about three-quarters of a mile or more below Lawn Lake. George Wagner oversaw the blasting activity. I remember only one day of blasting on the Pott's Puddle trail in 1980, with George using a couple of one-half, one-quarter, or one-eighth sticks of dynamite for the job.

Trail crews were in the Ranger Division when I started at Rocky Mountain National Park (RMNP), and Trails Foremen were badged and had law enforcement authority with minimal training. My first foreman, Mike Lynn, was very laid back about enforcement, and I took that attitude of communicating with visitors to heart, rather than writing tickets. Where enforcement was necessary, it was important, but enjoyment of the backcountry did not always need to be a strict enforcement message. I became a Trails Foreman in 1977.

As a result of increasing people and problems, the Park Service pushed its rangers to get more intensive law enforcement training. I decided that training in emergency medicine and rescue work was more to my liking. I was on the RMNP Rescue Team for four seasons, and trained first as an Emergency Medical Technician, and then as a Paramedic at Saint Anthony/Denver General Hospital programs. At Rocky, we often used Saint Anthony's Flight for Life Helicopters for rescue work. Their Alouette III helicopters worked well at high altitude, and their pilots and critical care flight nurses were superb.

While I was in paramedic school, I also worked for Western Camera in Estes Park, owned by Carl and Audre' Morris. I left Estes later in the fall of 1980, frustrated in an attempt to find full-time paramedic work in the mountains. I ended up in Grand Junction Colorado, where I met my future wife, Stephanie.

In early 1982, Stephanie had received a small inheritance from her grandfather, and we had the idea of purchasing the camera store from the Morris's. I knew that when I worked there, Carl was affiliated with the Western Camera store in Fort Collins, and the majority of his store inventory was on consignment from the Fort Collins store. Therefore, his inventory stock costs were low. When we approached them with a low offer of around $10,000, Carl told me that they had changed their operation, and now owned the inventory in the store. That put buying the camera store well out of our price range and killed our idea of moving back to the mountains at that time. While I was still living and working in Estes, they had moved from their old location on Elkhorn Avenue, adjacent to the alley near the Wheel Bar, to a west facing storefront just south of there on Elkhorn Avenue in the Park Center Mall. Fall River and the Big Thompson River ran through town in close proximity to the store and adjacent to most of downtown Estes Park.

We were living and working as paramedics in Colorado Springs in July 1982 when the Lawn Lake dam failed, and only saw the flood news in a peripheral sense. Several people had died, but not the numbers like the Big Thompson Flood in 1976. Estes Park was filled with mud, and we were happy we had not purchased the camera store. Our big family hike that summer had been into Thunder Lake in Wild Basin, which was early enough in the season that we were post-holing in snow for the last of the trail. Our teenage boys were excited about the fishing, and Stephanie was happy to have made the hike there and back.

In 1984, Stephanie and I were living on the western slope, in Silt Colorado, when I received a phone call from an attorney in Fort Collins. He had sent me a subpoena to testify in the Lawn Lake Flood case. I was asked to testify as to National Park Service Trail Crew blasting activity near Lawn Lake in the summer of 1981, prior to the flood. I told the attorney that I had last worked for the Park Service in 1980. He told me that I would still have to appear and testify in court in Fort Collins for a deposition to that effect. They would pay my travel costs for travel to Fort Collins and back to appear in court. I had to take a Greyhound bus for this testimony. The lawyer met me at the courthouse ahead of the hearing time and told me that they were trying to implicate the Park Service and George Wagner for the Lawn Lake Dam breach, causing the flood that killed three people. The lawyer was blaming trail crew blasting activity at or near Lawn Lake as the cause of the flood. George had since left the Park Service, and moved back east, according to the attorney. The gist of the court case according to this attorney, was that the Farmers Irrigation Ditch and Reservoir Company (FIDRC), which owned the dam and water rights in Lawn Lake, and was responsible for inspecting and maintaining the dam, and others, were trying to blame the National Park Service for the disaster. Since I didn't know who this lawyer was, I don't know if he was representing other claimants in this court case, or the ditch company that owned the dam.

When I was last at Lawn Lake in 1980, the Farmers Irrigation Ditch and Reservoir Company's employees, or some agency's inspectors were there "inspecting" the dam. They joked to me about doing their best to sample the fish population with fishing poles in inflatable rafts while they were there "inspecting". I saw little effort by these folks to look closely at the dam, which had been built starting in 1903. I thought that water leaked out of the dam into the Roaring River, just as it did at the old Arbuckle Reservoirs in Wild Basin, but I was wrong in my assumption. As an earthen dam, if it had leaked, it would have eroded and failed sooner. Water flow that I had seen coming out of the dam was likely coming from the gate valve being partially opened or flowing over the spillway. Since it was late August, I assume it was the outlet.

Our work in 1980 was drilling granite boulders on the Pott's Puddle Trail, a cutoff trail coming in from the northeast end of the Park. It joined the main Lawn Lake trail about three-quarters of a mile below the lake.

Where we were working was further down trail away from Lawn Lake. We were using Cobra gasoline-powered rock drills, which ran poorly, if at all, most of the time. Mostly we spent our days working on keeping the drills running, with little drilling accomplished after an hour plus hike in each day. I only recall the one day of blasting in 1980, with a few rocks removed from the trail. George Wagner was the East District Ranger and Trails Boss that year, and the licensed explosives expert. The other crew members I remember working on that section of trail were Pat Cleary, Jim Rabold, and Vern Groves. Pat, I had known for years, and Vern worked for me on my trail crew a few years before. Brady Wheeler was the foreman, and I had known him during my whole seasonal career with RMNP.

Before I gave my deposition in court, I also talked to George, who was wearing a western style suit jacket and cowboy boots that fit his former cowboy Park Service image. He told me he had retired and moved back east and was working with a church or bible camp. He seemed pretty nonchalant about what was going on at the court. My deposition was quick, I was asked if I worked on Trail Crew for the Park Service in 1981 or 1982 and I replied, "No", to all the questions, since I had left in 1980. I was dismissed and sat through a few minutes of the other testimony. Later I was mailed a copy of the court's transcription of my deposition to sign, and I signed it and mailed it back.

A few years later, I believe in 1985, living in Fort Collins, I recall a *Fort Collins Coloradoan* newspaper headline that the Park Service was found responsible in a court case relating to the Lawn Lake Flood. I didn't know which court case that was. I was surprised at that verdict.

When I was finishing my bachelor's degree at Colorado State University in 1988, one of the rangers that I knew in the seventies was in one of my classes. He was still with the Park Service. Frank Fiala told me he was working on a master's degree in water resources and law, and that Rocky Mountain National Park was removing the old dams in Wild Basin.

In 2003, I was on jury duty in Weld County, and the judge in the case asked me after the trial if I had testified in the Lawn Lake case years before. I replied that I had, and he told me he had been the Assistant U.S. Attorney for the Federal Government on that case. His name was Daniel Maus. I wanted to ask him how the case had resolved, but with his position as the judge, and me as a juror on a criminal case, I felt that wasn't something I could appropriately ask under the circumstances.

I have always been interested in dams and water in the west. My Grandfather, Herbert Prater, was the Chief Engineer at Region 7, for the Bureau of Reclamation (BOR) out of Denver, and during World War II, on active duty in the Army Corps of Engineers (ACE). He was involved in much of the Colorado-Big Thompson dam building projects and was working on the Glen Canyon Dam project before he retired. As a small child, my siblings and I hiked with granddad and grandma up the Wild Basin Trail to Calypso Cascades. That was one of my earliest recollections of the beauty and serenity that water played along the trails in that part of the Park. My grandfather gained his familiarity of this area when he was working on building the Olympus dam for Lake Estes and several of the western slope reservoirs of the Colorado-Big Thompson Project.

John Wesley Powell, the one-armed Union General from the Civil War, became the first Director of the U.S. Geological Survey. He wrote a monograph entitled *A Report on the Lands of the Arid Region of the United States, with a More Detailed Account of the Lands of Utah*, published in 1876. He was arguing that most of the west could not support farming without irrigation. Those lands that were adjacent to water were only about one to three percent of the west. He stated that dry land farming would require parcels of at least 2,560 acres, and those that were irrigated did not match the reality of the Homestead Act parcels of 160 acres. A further argument of Powell was that State boundaries did not respect the landscape, especially regarding drainages and watersheds, which are the natural water courses that should dictate boundaries.[2] Amazingly, his concept is utilized today, where ecosystem managers are more likely to use watershed boundaries for their studies, rather than arbitrary straight boundary lines drawn for convenience.

The homesteaders largely ignored Powell, believing the railroad companies' myths that "the rain will follow the plow", and moved west in great numbers to populate the arid lands. As we know from history, extended periods of drought ended many homesteaders' dreams.

Powell, also an adventurer, was part of the expedition as one of the first non-natives that successfully climbed Long's Peak in Rocky Mountain National Park in 1868. Much of his historical popularity was later based on accounts of his expeditions down the wild Colorado River in wooden boats.

When we lived in Idaho, there was much debate about the four dams on the Lower Snake River, and the loss of salmon upriver, as a result. This argument for de-damming is still going on today, but the tide is turning toward that being the best solution for salmon recovery.

At The Wildlife Society's 2012 Annual Meeting in Portland Oregon, I co-chaired a Native People's Wildlife Management Working Group Symposium on de-damming, presenting a paper entitled "A Brief History of Dam Building, De-damming, Restoration of Native Ecosystems and Their Benefits to Wildlife and Tribes."

When I had worked for the Shoshone-Paiute Tribes on the Duck Valley Indian Reservation, their fisheries management program was based on stocking three human- made reservoirs with trout as the Bonneville Power Administration's mitigation for the loss of native salmon runs due to damming the Owyhee River.

My interest in Rocky Mountain National Park's water resources has been largely a result of working there in the 1970's, and being familiar with the dams, lakes and streams there, as well as the other history of western water that I have acquired over the years. As part of trail crew,

I built many small bridges and a few larger bridges on Wild Basin trails. I was in Estes Park during the Big Thompson Flood of 1976 and living on the Front Range of Colorado during the floods of 2013, both caused by stalled natural storm events. More frequent catastrophic weather events, likely due to human-caused climate disruption[3,4], and the risks of aging dams require us to be more proactive in our mitigation efforts. Recent droughts and the resulting large wildfires in the west add to the burden of keeping our watersheds intact, water supplies safe, and flooding risks reduced. Rocky Mountain National Park had two large forest fires burning outside the Park in 2020, that ended up burning well into the Park before the season ended. As predicted by Powell and others, fires and watersheds don't respect straight line boundaries.

In the summer season of 2021, mud slides and flooding have occurred in the 2020 burn sites throughout Colorado. The Grizzly Creek burn area in Glenwood Canyon has closed Interstate-70 highway numerous times this season. The Cameron Peak burn area north of Rocky Mountain National Park, caused mudslides and flash flooding on the upper Poudre River in July, causing two deaths, with three more missing and presumed dead family members as of this writing.

I originally started this book as a look at the Lawn Lake flood of 1982 because I was asked to testify about it and decided that the larger picture really included the water resources of Rocky Mountain National Park and adjacent areas, and the history of the human need to control and harness these water resources for use outside the Park. Whether located in the Park, or outside the Park, water influences decisions that concern the management of RMNP and its surrounding region.

Rocky Mountain National Park is truly an urban park, it is the playground for a rapidly growing Colorado Front Range population, whose water needs outside the Park heavily influence regional water politics. The human impact of the Front Range growth will be the beast that needs eventual taming, if the Park is to survive.

# CHAPTER ONE

# Water Resources in Rocky Mountain National Park

R ocky Mountain National Park is split by the Continental Divide, where water normally flows west towards the Sea of Cortez into the Pacific Ocean, or water flows east toward the Mississippi River and the Gulf of Mexico. Prior to the establishment of the Park, water hungry farmers, through their ditch and reservoir companies, started diverting waters to flow east. These diversions are discussed in later chapters.

The majority of the Park lies east of the 42-mile-long divide, so most water naturally flows east already. The average east side precipitation is 16.8 inches of water, while the west side precipitation averages 19.9 inches of moisture, annually.[1]

This water is a precious resource to the plant and animal communities of Rocky Mountain National Park, and an important part of the visitor experience, whether that be as ice, snow, creeks, waterfalls, rivers or lakes.

This chapter is intended to describe the naturally occurring flow of melting ice and snow into creeks and rivers and lakes in the Park, divided by the individual drainages as they come down the mountains and eventually flow out of the Park. These are the waters that sustain the ecosystems and provide enjoyment for visitors of Rocky Mountain National Park, before they come under use by water managers outside the Park. The Park Information Office lists a total of 147 lakes in the Park, covering 1151 acres, or 466 hectares.[1]

Five rivers eventually drain most Park waters. These are the Cache La Poudre River, Fall River, the Big Thompson River, The North Saint Vrain River, and the Colorado River.

My descriptions will flow as the water does, from higher mountain slopes downhill and downstream as they converge with other streams and rivers. Trails located near these streams will often be included with these drainage descriptions. Of the 355 miles of trails in the Park[1], many of them intersect or follow drainages as they lead to many of these lake destinations. Hiking a trail with lakes and streams along its way is far more satisfying than hiking the "dry" trails that are far fewer in number in this Park. Susan Joy Paul describes 43 named waterfalls in Rocky Mountain National Park in *Hiking Waterfalls in Colorado*, including detailed descriptions and hiking directions.[2] I have tried to identify these waterfalls in my descriptions to match her names, so they can be found more easily. Falls with unofficial names are in quotations, such as "Falcon Falls".

There are many trail guides written about Rocky Mountain National Park. While I have described trails adjacent to waterways, details about specific trails and trailheads may be found in these books or on the internet. Former rangers Kent and Donna Dannen wrote several popular guides to hiking trails, including *Rocky Mountain National Park Hiking Trails*, the Third Revised Edition is referenced here.[3]

There are 25 topographical maps (quadrangles) covering Rocky Mountain National Park and its adjacent areas, 1:24000-scale, 7.5-minute maps. To the left is an index to those maps covering the Park.[4]

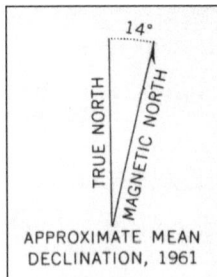

| CLARK PEAK | CHAMBERS LAKE | COMANCHE PEAK | PINGREE PARK | CRYSTAL MOUNTAIN |
|---|---|---|---|---|
| MOUNT RICHTHOFEN 1957 | FALL RIVER PASS 1958 | TRAIL RIDGE 1957 | ESTES PARK 1957 | GLEN HAVEN |
| BOWEN MTN. 1957 | GRAND LAKE 1958 | MCHENRYS PEAK 1957 | LONGS PEAK 1957 | PANORAMA PEAK |
| TRAIL MTN. 1957 | SHADOW MOUNTAIN 1958 | ISOLATION PEAK 1958 | ALLENS PARK 1957 | RAYMOND 1957 |
| GRANBY 1957 | STRAWBERRY LAKE 1958 | MONARCH LAKE 1958 | WARD 1957 | GOLD HILL 1957 |

14°

TRUE NORTH

MAGNETIC NORTH

APPROXIMATE MEAN DECLINATION, 1961

**7.5-minute topographical maps for RMNP and adjacent areas:**

| | | NORTH | | | | |
|---|---|---|---|---|---|---|
| | Clark Peak | Chambers Lake | Comanche Peak | Pingree Park | Crystal Mtn. | |
| | Mt. Richthofen | Fall River Pass | Trail Ridge | Estes Park | Glenhaven | |
| WEST | Bowen Mtn. | Grand Lake | McHenrys Peak | Longs Peak | Panorama Peak | EAST |
| | Trail Mtn. | Shadow Mtn. | Isolation Peak | Allens Park | Raymond | |
| | Granby | Strawberry Lake | Monarch Lake | Ward | Gold Hill | |
| | | SOUTH | | | | |

# Northeast: Trail Ridge, Comanche Peak, Estes Park, Pingree Park, Crystal Mountain, Glenhaven Quadrangles

The South Fork of the Cache La Poudre River flows off Mummy Pass and Flint Pass between Fall Mountain and Ramsey Peak out of the northcentral Park boundary. The Stormy Peaks Trail and Mummy Pass Trail lie on either side of this drainage. A creek coming off north, between Rowe Peak and Rowe Mountain to Icefield Pass, also flows into this drainage. A couple of small creeks come off the north side of Stormy Peaks before flowing into the South Fork of the Cache La Poudre River outside the Park.

Fall Creek flows north off Fall Mountain in two branches, one originating in the Park, and the second out of a series of small lakes north of the Park, including Cirque Lake, between Fall Mountain and Comanche Peak. Three glaciers (permanent snowfields) line the ridge between Fall Mountain and Comanche Peak.

A large cirque is formed east of the Mummy Range; defined by Stormy Peaks to the northeast, Sugarloaf to the north, Skull Point to the northwest, Rowe Mountain, Rowe Peak, and Hagues Peak to the west, and Mummy Mountain, Mount Dunraven and Mount Dickenson to the south.

Out of this cirque flow waters off Rowe Peak, beginning with Rowe Glacier flowing through several unnamed lakes to Lake Dunraven, then into the North Fork of the Big Thompson River below Lake Louise, which drains Rowe Mountain, Icefield Pass and Skull Point. Lake Louise drains through a few small ponds before it joins and becomes the North Fork of the Big Thompson River. Isolated Lake Husted lies between Lake Louise and Lost Lake. Lost Lake flows into the North Fork, and an unnamed stream runs off the northeast side of Mount Dunraven and joins along the Lost Lake Trail.

Flowing north off Sugarloaf Mountain and Ramsey Peak is a tributary of the South Fork of the Poudre River, which heads out of the Park as well.

The Lost Lake Trail parallels much of the river and then forks near Lost Falls, with the Stormy Peaks Trail going northwest, and the Signal Mountain Trail going east. A small drainage joins the North Fork below Lost Falls and the river continues to parallel the Lost Lake trail going east to the North Fork Ranger Station. The trail splits near the ranger station, into the Dunraven Trail heading east, and the North Boundary Trail, heading southeast. An unnamed creek comes off the east side of Mount Dickinson and joins the North Fork after it leaves the Park boundary along the Dunraven Trail.

Five unnamed creeks drain off the east side of Stormy Peaks and off South Signal Mountain, all flowing to the north out of the Park, or joining Pennock Creek, before running into the South Fork of the Cache La Poudre River. The Signal Mountain trail circles around much of that area in the northeast corner of the Park.

Unnamed streams south of these trails come off the east side of Mount Dickinson. Off the east side of Mummy Mountain and Mount Dunraven flows West Creek joined by other unnamed creeks flowing down West Falls before joining the North Boundary Trail below the falls outside the east Park boundary. Cow Creek joins West Creek before flowing into Devil's Gulch. Grouse Creek and Fox Creek drain off the east side of Mount Dickenson, with Fox Creek Falls along the trail just outside the Park boundary, before joining the North Fork of the Big Thompson near Glen Haven. The North Boundary Trail comes into the McGraw Ranch. Also, out of the McGraw Ranch is the Cow Creek trail following Cow Creek up to Bridal Veil Falls. The trail splits and becomes the Black Canyon Trail between MacGregor Mountain and Dark Mountain to the North. The Black Canyon Trail goes past Potts Puddle and joins the Lawn Lake Trail, about three-quarters of a mile below Lawn Lake. Black Canyon Creek comes off the southwest side of Mummy Mountain, between Lumpy Ridge to the east and MacGregor Mountain to the west. Mount Tileston and Bighorn Mountain separate these drainages from the Roaring River coming out of Lawn Lake. Black Canyon Creek eventually flows between McGregor Mountain and Lumpy Ridge, south on Devils Gulch into the Big Thompson River before Lake Estes.

An isolated puddle called Gem Lake is found off the east end of Lumpy Ridge and Twin Owls rock formation. Gem Lake may be hiked to from the Twin Owls parking lot, the MacGregor Ranch Road, or from a third location on Devils Gulch Road.[3] Although the lake is just a small, shallow pond, it is ringed by worn rock formations. I used to hike the two miles in, rock climb and traverse the cliffs, and hike back out for a daily exercise routine.

The Roaring River comes off to the east of the Saddle between Hagues Peak and Fairchild Mountain. It joins water out of Crystal Lake and Lawn Lake heading south between Fairchild and Mount Tileston. A small creek drains a few ponds and joins the Roaring River west of Mount Tileston and then continues south on the west side of Bighorn Mountain.

Between Fairchild Mountain and Mount Ypsilon, a creek drains into the Fay Lakes, before joining Ypsilon Creek to the east of Ypsilon Lake. Spectacle Lakes come off the northeast ridge of Mount Ypsilon and drain into Ypsilon Lake. Chiquita Lake, coming off a cirque on the eastern slope of Mount Chiquita, also drains into Ypsilon Lake. Ypsilon Creek drains to the southeast before joining the Roaring River to the west of Bighorn Mountain. The beautiful, isolated Chipmunk Lake sits on a flat spot southeast of Ypsilon Lake, about 4 miles in on the trail to Lake Ypsilon. Finally, Roaring River drops down between the slope of Bighorn and Mount Chiquita over Horseshoe Falls and enters Horseshoe Park and the Fall River. These dramatic falls became more spectacular as uprooted trees and large boulders came down the hillside from the Lawn Lake flood, creating a large alluvial fan where it entered the valley.

*The Cache La Poudre River Headwaters as seen from Trail Ridge Road.*

Chiquita Creek comes off the slopes between Mount Chiquita and Mount Chapin and an unnamed creek joins it before draining across Old Fall River Road into Fall River near the Endovalley picnic area.

In steep cirques defined by Desolation Peaks, Mount Ypsilon, and Fairchild Mountain, drains Hague Creek flowing north, joined by two creeks coming off unnamed lakes. Hague Creek is joined by an unnamed creek coming off The Saddle between Fairchild and Hagues Peaks. Mummy Pass Creek flows off the western flank of Mummy Pass and enters Hague Creek. A creek coming off the eastern flank of Flatiron Mountain, flows out of four small lakes into Hazeline Lake before continuing north into Hague Creek. Hague Creek continues to the northwest, joined by one more unnamed creek before eventually

*Unnamed Falls off Old Fall River Road.*

flowing into the Cache La Poudre (Poudre) river. The Mummy Pass Trail comes down from the pass and joins the trail along the Poudre River.

Chapin Creek drains the west side of Mount Chapin, Chiquita, Ypsilon, Desolation and Flatiron Mountains and enters the Poudre River. Cascade Creek comes out of Mirror Lake and drains to the west into the Poudre River. There is a trail from the Poudre River to Mirror Lake. Mirror Lake itself sits outside the north boundary of Rocky Mountain National Park.

From the west, between Chapin and Hague Creeks, out of Long Draw Reservoir, La Poudre Pass Creek enters the Poudre River.

# Fall River Pass, Trail Ridge, Estes Park Quadrangles

Fall River originates near Fall River Pass, by the Alpine Visitor Center, drains through Willow Park, collects water cascading down from Iceberg Lake and follows much of the Old Fall River Road. Beautiful Chasm Falls is just off the road 1.4 miles up from the junction at Endovalley Picnic Area. Sundance Creek comes off the eastern flank of Trail Ridge, south of Sundance Mountain, flows through Hanging Valley, and then drops down Thousand Falls and enters Fall River at Endovalley. Fall River meanders through Horseshoe Park before leaving the National Park through Cascade Lake, Aspenglen Campground and the Fall River Entrance.

Hidden Valley Creek originates off the northeast side of Tombstone Ridge, before crossing Trail Ridge Road, and draining through Hidden Valley (former ski area) and flowing through an area of beaver ponds near Trail Ridge Road. Hidden Valley Creek joins Fall River in Horseshoe Park above Cascade Lake. Bighorn Creek flows off the east side of Bighorn Mountain and crosses the highway near the entrance station before joining Fall River above the Power Plant and State Fish Hatchery. A spring also comes off the south side of Bighorn Mountain north of isolated Sheep Lakes and flows through Hondius Park before joining Fall River. Fall River enters the Town of Estes Park and merges with the Big Thompson River in the downtown area before they flow into Lake Estes.

# North of Wild Basin-East Side-Long's Peak area:
## Long's Peak, McHenry's Peak Quadrangles

Horse Creek flows off the north slopes of Lookout Mountain and Horsetooth Mountain to the east into the Meeker Park area at Colorado Highway 7, before it joins the North Saint Vrain River.

Cabin Creek drains the eastern slope of Mount Meeker and comes down to Camp St. Malo near Highway 7. Two other unnamed creeks come off and join from the eastern side of Meeker to the north of Camp St. Malo and flow into Cabin Creek at Camp St. Malo. The East Face of Long's Peak, Lamb's Slide, Glacier Ridge, and Mill's Glacier, drain into Chasm Lake down through Chasm Meadows and into Peacock Pool. Drainage off Ship's Prow and the north face of Mount Meeker also flow into Peacock Pool.

*Peacock Pool (left) and Columbine Falls (right), Long's Peak Area.*

This is Roaring Fork Creek from Peacock Pool that drains into Tahosa Creek on the east side of Highway 7, and finally into the North St. Vrain River.

Larkspur Creek and waters coming off of Jim's Grove drain the eastern side of Battle Mountain off the south side of Pine Ridge. Hiking up the mountain, it is the first creek to cross the East Long's Peak trail before joining Alpine Brook that flows on down the mountain. After crossing Larkspur creek going up the trail, Alpine Brook tumbles beautifully across the trail just before tree line. Downhill after flowing near the Long's Peak Ranger Station trailhead, Alpine Brook eventually joins Tahosa creek on the east side of Highway 7.

South of Storm Pass and north of Pine Ridge off Battle Mountain, flows Inn Brook past Eugenia Mine down past the old Long's Peak Inn. A series of ponds and a creek in the Tahosa Valley join Inn Brook before it enters Tahosa Creek.

Wind River creek flows off the north side of Battle Mountain to the East Portal of the Alva B. Adams Tunnel. Wind River and Aspen Brook flow into the Big Thompson River, but the Aspen Creek Siphon takes water from the East Portal to the Rams Horn Tunnel (under Rams Horn Mountain) to Mary's Lake. From there it is pumped through Prospect Mountain into Lake Estes.

Northeast of Estes Cone, off Wind River Pass, flows Aspen Brook on the back side of Lily Mountain, draining west of Rams Horn Mountain to meet the Big Thompson River. Lily Lake drains across Highway 7 into Fish Creek, which flows down the North Side of the Crags, joined by a small stream near the Cheley Camps, before flowing into the southern spur of Lake Estes, approximately five miles downstream.

Flowing out of the trickling rocks of the Boulder Field, and Mount Lady Washington, off the north side of Long's Peak, is the beginning of Boulder Brook. It is joined by an unnamed creek coming off the north side of Storm Peak. Boulder Brook runs down past the North Long's

Peak Trail and eventually drains into a series of beaver ponds in Glacier Basin, and into Sprague Lake. Boulder Brook merges into Glacier Creek before the Glacier Basin Campground area.

Off of the west side of Long's Peak, in cirques formed by Storm Peak, Long's, Pagoda, Chief's Head, The Spearhead, and McHenry's Peaks, are a series of high lakes and streams. Draining from snowfields on the northeast side of Chief's Head, small tarns and Green Lake flow into Black Lake. Blue Lake comes off the west side of Storm Peak and drains into Glacier Creek below Black Lake. Between The Spearhead and Stone Man's Pass, Frozen Lake flows into Black Lake. Below Black Lake is Ribbon Falls on Glacier Creek, joined by Shelf Creek flowing off Powell Peak (Powell snowfield or glacier) into Solitude Lake. Shelf Lakes also drain into Shelf Creek and then into Glacier Creek between Arrowhead and

*Alberta Falls in Winter, Glacier Gorge Trail.*

Thatchtop Mountain. Glacier Creek continues downstream into Jewel Lake and Mills Lake in Glacier Gorge. These high lakes and creeks in Glacier Gorge are some of the most spectacular views in the Park. Hiking uphill, after Alberta Falls, Glacier Gorge Trail branches early and follows up with the east (left) branch past Glacier Falls, Mills Lake and Black Lake. Scrambling across rocks takes you past Black Lake to Jewel Lake and Green Lake, an unnamed tarn above Green Lake has been informally named "Italy Lake" in the Dannen book.[3]

Icy Brook flows off of Taylor Glacier, between Powell Peak and Taylor Mountain, flowing down Icy Loch Vale into Sky Pond, Lake of Glass, over Timberline Falls and into The Loch. Andrews Creek coming off Andrews Glacier south of Otis Peak, drains into Icy Brook just above The Loch. Hiking uphill, the west (right) branch of the Glacier Gorge Trail past Alberta Falls, goes to The Loch, Timberline Falls, Lake of Glass and Sky Pond. A tiny lake, Embryo Pond lies on the west side of the base of Thatchtop Mountain, just below Timberline Falls on Andrews Creek. The North Long's Peak Trail comes off at a junction above Alberta Falls before the junction of Loch Vale and the Mills Lake Trail.

Just a note about the Alva B. Adams tunnel. This water diversion tunnel is bored from southwest to the northeast under Rocky Mountain National Park, from the West Portal at Grand Lake, and passes under Andrews Glacier just south of Otis Peak at the Continental Divide. It empties at the East Portal into the Wind River that drains into Aspen Brook. The Alva B. Adams tunnel will be discussed in another chapter.

Between Otis Peak and Hallett Peak is Otis Gorge. Chaos Creek flows off the snowfields (glaciers) into Lake Haiyaha, also draining the Glacier Knobs area. Chaos Canyon Cascades are a waterfall on Chaos Creek before it joins Glacier Creek near Bear Lake Road.

Tyndall Gorge lies between Hallett Peak and Flattop Mountain, and Tyndall Glacier drains into Tyndall Creek into Emerald Lake, and

then Dream Lake. Tyndall Creek then joins Chaos Creek. Nymph Lake
is isolated between Dream Lake and Bear Lake. Bear Lake drains into
Chaos Creek before it joins Glacier Creek. There are trails from the Bear
Lake area to Nymph, Dream and Emerald Lakes, the Flattop Mountain
trail forks off to Flattop, and up to the Continental Divide. A trail junction
near Dream Lake takes you to Lake Haiyaha, with other trails joining
the Glacier Basin trail network as well.

Out of Odessa Gorge, between Flattop and Notchtop mountains,
flow waters off Ptarmigan Point and the snowfield off Ptarmigan Pass.
From a number of unnamed ponds, cascading down Grace Falls, the
creek is joined by waters from Lake Helene before flowing into Odessa
Lake. North of Knobtop Mountain and the Little Matterhorn, waters
from snowfields cascade down Tourmaline Gorge and Tourmaline Lake

*Fern Falls*

into Odessa Lake. These flow past The Gable into Fern Lake. Fern Lake waters drop over Marguerite falls as Fern Creek, then Fern Falls, before joining Spruce Creek above The Pool.

The Fern Lake Trail continues past Fern Lake to Odessa Lake and Gorge, up Joe Mills Mountain, traverses adjacent to Marigold Pond, before heading either to Bear Lake off the Flattop trail, or Bierstadt Lake or Mill Creek Basin.

Adjacent to Lake Helene, Two Rivers Lake and Marigold Pond drain off Joe Mills Mountain as Mill Creek, joined by an unnamed creek off the northeast side of Flattop Mountain before flowing into Mill Creek Basin. Mill Creek is joined by waters draining out of Bierstadt Lake, before flowing into Hallowell Park, flanked by the Bierstadt Moraine on the south and the South Lateral Moraine north of Steep Mountain. Mill Creek flows into the Big Thompson River coming out of Moraine Park.

Coming off Sprague Pass and Bighorn Flats on the Continental Divide, Spruce Creek is joined by waters from Sprague Glacier, Irene Lake and Rainbow Lake off Sprague Mountain and Stone Peak. Hourglass Lake also drains into Spruce Creek high off that cirque. Hidden River flows from the east off Stone Peak, and then Loomis Lake and Spruce Lake join Spruce Creek off Gabletop Mountain.

Fern Creek and Spruce Creek merge and join the Big Thompson River coming out of Forest Canyon. Below that is The Pool along the Fern Lake Trail. A small creek on the north side of Mount Wuh, flows into those merging waters at the Big Thompson below The Pool. Cub Lake, isolated southeast of Arch Rocks, drains into a series of small ponds and beaver ponds, before joining the Big Thompson River in Moraine Park.

# Central: Fall River Pass, Trail Ridge, Grand Lake, McHenry's Peak, Long's Peak Quadrangles

The Big Thompson River originates off Forest Canyon Pass, east of the Continental Divide near Milner Pass, and south of Trail Ridge Road. Two unnamed creeks flow off of the divide, between Sheep Rock and Mount Ida. The southernmost of these two drains a small cirque with a series of small lakes north of the ridge coming off Mount Ida. The Gorge Lakes originate between Mount Ida and Cracktop Mountain, draining the cirque formed north of Cracktop, Mount Julian and Terra Tomah Mountains. Highest Lake drains off the divide into Azure Lake, then into Inkwell Lake, flowing into Arrowhead Lake. Isolated Lone Lake sits north of Arrowhead Lake, and Doughnut Lake flows into the creek out of Arrowhead, before joining Rock Lake, Little Rock Lake and Forest Lake, before joining the Big Thompson River. The Gorge Lakes are spectacular

*Mount Ida and the Gorge Lakes from Trail Ridge Road.*

from any view, whether from the Continental Divide, or viewed from Trail Ridge Road. The easiest hike to Mount Ida is to follow the divide south from Milner Pass by Poudre Lake.

South of the massif formed by Cracktop, Julian, and Terra Tomah mountains, is the Hayden Gorge, originating north of Hayden's Spire, draining the cirque and valley formed by Hayden's Spire, Sprague Mountain and Stones Peak. Hayden Lake and Lonesome Lake, and unnamed ponds and creeks flow as Hayden Creek before joining the Big Thompson River in Forest Canyon. Three unnamed streams flow off the south face of the canyon below Rock Cut, and Forest Canyon Overlook below Trail Ridge Road. From the northeast side of Stone Peak, flows Lost Brook into Raspberry Park, before the Big Thompson River joins the drainages out of Spruce Canyon. Beaver Brook comes off the northeast side of Beaver Mountain and flows out of Beaver Meadows and past the entrance station, north of Eagle Cliff Mountain, before joining the Big Thompson River. Buck Creek comes off the south side of Deer Mountain and merges into Beaver Brook.

All of the merging creeks to the north and west of Long's Peak on the east side of the Continental Divide, eventually join the Big Thompson River, including Glacier Creek, Aspen Creek, Mill Creek, Fern Creek and Spruce Creek. Fern and Spruce Creek join before leaving the boundary of the Park. Mill Creek, Glacier Creek, Aspen Brook, and Wind River join east of Tuxedo Park, outside the Park boundary.

# Southeast: Wild Basin-Isolation Peak, Allens Park Quadrangles

## South to North

South of Mount Copeland, and east of Ogallala Peak lie the Cony and Hutcheson Lakes, along the southeastern boundary of the Park. A difficult hike, Cony Lake and the Hutcheson Lakes drain north of Elk Tooth down Cony Creek near Pear Lake (formerly Pear Reservoir). Pear Lake drains into Cony Creek about one-half mile below the lake. Cony Creek runs adjacent to Finch Lake that drains a small outlet creek into Cony. A small creek comes down from between Saint Vrain Mountain and Meadow Mountain and joins Cony Creek below Finch Lake. This and one other small creek coming off Meadow Mountain are the only water relief on the otherwise dry trail on the hike to Finch Lake. Cony Creek then plummets down the hillside to form Calypso Cascades, a popular hike just 1.8 miles up the main Wild Basin trail. Cony then joins the North Saint Vrain Creek as it heads east along the main trail. As you hike from the Wild Basin Trailhead, Copeland Falls and Upper Copeland Falls are found one-third of a mile up the trail, much of the trail follows the North St. Vrain Creek, with the next "falls" just up trail from the CCC bridge, called "Lover's Leap", by Susan Paul[2], about 1.5 miles up the main trail.

*Bull moose near Chickadee Pond, off Bluebird Lake Trail.*

Willow Creek flows off the northeast slopes of Meadow Mountain into Allenspark. This is likely the source of the Allenspark Spring. North of Mount Copeland, Junco Lake flows from the cirque of Ouzel Creek, Cony Pass and Mount Copeland and becomes part of Ouzel creek after it joins with the drainage of Pipit Lake and Lark Pond that feed into Bluebird Lake (formerly reservoir), which drain the next cirque north, bounded by Mahana Peak to the north. On a small flat on the west side of Mahana, lies Isolation Lake. To the west and north is Isolation Peak. Bluebird drains into Ouzel Creek, which feeds into Ouzel Lake. A small pond covered with Yellow Pond Lilies, Chickadee Pond, sits between Ouzel Lake and the Bluebird Lake trail. This pond is frequented by moose, seen by the author here in the 1970's and again in 2014.

Ouzel Creek plunges off a cliff in a beautiful scenic drop as Ouzel Falls, before flowing down the hillside to join the North St. Vrain Creek.

*Ouzel Falls-Wild Basin Area.*

Ouzel Falls is three miles up the main Wild Basin Trail from the ranger station.

Between Mahana Peak and Tanima Peak lies a high cirque below Moomaw Glacier, where Frigid Lake and tiny Indigo Pond drain into Eagle Lake. Box Lake, just to the north and Eagle drain off the high plateau down exquisitely beautiful Mertensia Falls, so named for the bluebells cascading along the falls. Mertensia drains into the North Saint Vrain Creek coming out of the Thunder Lake area. Thunder Lake, along the northeastern flank of Tanima Peak, fills from the cirque bounded by Tanima, Boulder-Grand Pass, and Pilot Mountain. Two lakes, Lake of Many Winds, and Falcon Lake drain into Thunder Lake. "Falcon Falls"[2] pours out of Falcon Lake on its way down the mountainside to Thunder. A creek coming off a ridge north of Mount Alice appears to be the source of the North Saint Vrain, rather than the creek coming out of Thunder Lake. An unnamed pond on this stream, and the unnamed pond above Fan Falls, on the tundra plateau between Pilot Mountain and Mount Alice were given names, Christine and Alice, by the author. Following the creek adjacent to the Thunder Lake Patrol Cabin, uphill to the plateau is the easiest access to this beautiful plateau area to the north.On the tundra plateau to the north of Thunder Lake, below the North Ridge of Mount Orton, Snowbank Lake comes off the slopes of Mount Alice, and drains into Lion Lake 2. Lion Lake 2 continues flowing down to Trio Falls that drains into Lion Lake 1. Trio Falls has three or more plumes trickling off the rock face into a snowbank. This creek then comes down Thunder Falls and joins the North St. Vrain below Thunder Lake. There is an improved trail cutoff up the hill 4.7 miles in, on the main trail to Thunder Lake, to this series of lakes. The trail is improved to Lion Lake 1, but not to the upper two lakes. Snowbank Lake is one of the original "Arbuckle Reservoirs", which also included Pear, Bluebird and Sandbeach Lakes.

To the east off the trail to Lion Lake 1, off the southwest side of the North Ridge of Mount Orton, is a small, isolated lake named Castle Lake. Two small ponds, named Castle Lakes, lie approximately six-tenths of a mile south of Castle Lake. Out of the cirque south of Pagoda Mountain, Keplinger Lake drains down as Hunter's Creek, a few unnamed ponds along the way. Other unnamed creeks drain off Dragons Egg Rock and the south side of Mount Meeker to join Hunter's Creek. Hunter's Creek

*Snowbank Lake, Wild Basin.*

continues downhill with Lyric Falls and drains into the North St. Vrain near the Wild Basin Trailhead and ranger station. On the 4.5-mile trail to Sandbeach Lake (formerly reservoir), Hunter's is the upper creek crossing on the trail to Sandbeach Lake. Sandbeach Lake collects melted snow and water off the eastern slopes of Mount Orton and drains out Sandbeach Creek into the North Saint Vrain Creek above Copeland Falls on the main Wild Basin trail. Two small, isolated lakes, called Twin Lakes, lie on the North Ridge between Sandbeach Creek and the main North St. Vrain valley. Camper's Creek drains off Horsetooth and Lookout Mountain to the east of Hunter's Creek and drops off the Copeland Moraine into North Saint Vrain Creek. Camper's creek is the first water crossed on the hike up the moraine on the way to Sandbeach Lake. Both Camper's Creek and Hunter's Creek provide relief on the hike up the steep moraine on the way to Sandbeach Lake. At the entrance to Wild Basin, Copeland Lake (Reservoir) is filled off water diverted from the North Saint Vrain. Copeland Lake is owned by the Saint Vrain and Left Hand Water Conservation District (SVLHWCD) for the City of Longmont, but co-managed with the National Park Service. The water level in the lake is kept low due to the hazardous condition of the dam.

# Southwest side Lakes and Streams: Isolation Peak, Shadow Mountain Quadrangles

Starting in the south at the Continental Divide, the west side of the Park is bounded by Ogalalla Peak at the divide, and Watanga Mountain and Mount Adams to the west. Watanga creek flows off the southwest flank of Watanga Mountain out of the Park. The large drainage of Paradise Park begins with drainage off of Ogalalla, Ouzel and Isolation Peaks, and is joined by waters draining out of Adams Lake off the northeastern flank of Mount Adams. Paradise Creek flows northwesterly before it joins the East Inlet drainage. The easiest access from the west side of the Park is up the East Inlet trail, and then follow Paradise Creek into this area. A short trail starts south along the western edge of Mount Craig, but then it is bushwhacking along the creek the rest of the way. A large number of small unnamed lakes dot the alpine tundra between Isolation peak and Mount Craig, which forms a divide between Paradise Park and the lakes of the East Inlet Trail. Some larger isolated lakes sit above Ten Lake Park, between Mount Craig and Paradise Park. Paradise Park valley south of Mount Craig and Isolation Peak is a Research Natural Area, with no maintained trails and no camping allowed. This is one of the most beautiful areas in RMNP, with only unnamed lakes except for Adams Lake. High country fishing is the most common activity here.

West of Mount Adams flows the multi-branched Columbine Creek, draining the area between Mount Acoma and Twin Peaks to the south, and Mount Bryant to the north. Columbine Creek drains into Columbine Bay, on the Colorado River between Shadow Mountain Lake and Lake Granby. The Columbine Creek Trail is south of the creek, joining the creek near trails end, probably more of a fishing spot than a day hike destination. West of Mount Bryant, is Pole Creek, draining into the Colorado River as well, between Shadow Mountain Lake and Lake Granby. Off the southwest side of Shadow Mountain, flows Ranger Creek, coming into the Colorado River below the Shadow Mountain Dam. The Shadow Mountain Trail comes off the Outlet Trail between Shadow Mountain Lake and the Colorado River, the East Shore Dam Trailhead is located at the south end of Shadow Mountain Lake. The Shadow Mountain Lookout is the high point of this area.

To the west of the divide across from Wild Basin, and north of Paradise Park, flow the waters coming off the north of Isolation Peak to Fifth Lake down the East Inlet Creek into Fourth Lake. Fourth Lake is just west of Boulder-Grand Pass on the Continental Divide. This is one of the most beautiful trips to make across the divide, from Thunder Lake over Boulder-Grand Pass and down the East Inlet trail, or the reverse trip from west to east. Fourth Lake flows into Spirit Lake and into Lake Verna. Three unnamed streams flow off the north slopes between Isolation Peak and Mount Craig, joining the East Inlet creek, originally coming out of Fifth Lake. Lone Pine Lake represents the first (lower) lake of the five major lakes that make up this drainage. A falls about 0.75 miles below Lone Pine Lake cascades down between Mount Cairns and Mount Wescott, below the Paradise Creek Junction with East Inlet. Echo Creek flows out of the drainage formed between Mount Wescott to the west and Mount Bryant to the east, into the East Inlet west of Adams Falls. Adams Falls is a small detour off the East Inlet Trail, about 0.6 miles (mi.) from the East Inlet Trailhead. The Falls is a cascade of 50-60 feet down a rocky slope and is quite spectacular for a short hike. The East Inlet flows into the eastern shore of Grand Lake, just south of the West Portal of the Alva B. Adams Tunnel.

## West Central: Grand Lake, McHenry's Peak, Fall River Pass Quadrangles

The North Inlet (of Grand Lake) flows off the Continental Divide from Lake Powell off McHenry's Peak and Powell Peak. From Lake Powell, and a small creek off the north side of Mount Alice, this drainage begins and flows to the northwest. It gathers water from a series of small lakes off the east side of Andrews Peak, joined by flows off the north side of Andrews, which flow into Lake Nanita. Lake Nanita flows into the North Inlet above Lake Solitude (not Solitude Lake-east of the divide), which is joined by waters off Taylor Peak and Lake Nokoni, flowing off Ptarmigan Mountain. The North Inlet Trail switchbacks up to Lake Nokoni and then to Lake Nanita. Below Nokoni, at Hallett Creek, the trail splits off the North Inlet Trail onto the Continental Divide Trail, heading up to the flats on the divide.

Hallett Creek flows from a drainage west off Andrews Pass and Hallett Peak. Out of a high cirque off Ptarmigan Point to the northeast, and Snowdrift Peak to the west, flow five smaller creeks into Ptarmigan Creek. Ptarmigan Creek comes out of Ptarmigan Lake, an unnamed lake, and Snowdrift Lake. These waters are joined by two other creeks into Bench Lake, then flow over War Dance Falls, before joining the North Inlet Creek. As the creek and trail continue downstream, they are joined by five other creeks flowing off Mount Patterson, Nisa Mountain, and Ptarmigan Mountain. Lake Pettingell drains off the northwest side of Ptarmigan Mountain. The North Inlet Creek then tumbles over Cascade Falls, before being joined by waters off Mount Enentah, and another stream off the south side of Nisa Mountain. The North Inlet flows through Summerland Park and all these joined waters then drop into Grand Lake.

Susan Paul describes seven waterfalls on the North Inlet Trail, some with her unofficial names (in quotation marks). Hiking uphill the first falls is Cascade Falls (3.5 mi.), "Snake Dance Falls" (4.5 mi.), "Big Pool Falls" (4.9 mi.), "Sun Dance Falls" (5 mi.), "Rain Dance Falls" (5.7 mi.), War Dance Falls (7 mi.), North Inlet Falls (7.7 mi.).[2]

The remnant of Eureka Ditch off Sprague Pass, coming across Bighorn Flats, is a historical diversion that can still be seen along the Continental Divide. This ditch was dug to divert water from flowing west off Bighorn Flats, to go east to Spruce Canyon to flow down Spruce Creek.

The Flattop Mountain Trail joins along the Continental Divide Trail with the Tonahutu Creek Trail described in the next drainage. Off the north side of Snowdrift Peak, drains glacial waters into Murphy Lake, joined by six small creeks draining the Bighorn Flats and Sprague Mountain. One of the creeks drains the Haynach Lakes coming off the divide and Nakai Peak, to form Tonahutu Creek. These waters flow over Granite Falls and are joined by more waters of the south side of Nakai Peak. Tonahutu

Creek continues south through Big Meadows toward Grand Lake, flanked by Green Mountain to the west and Mount Patterson to the east. Nisa Mountain adds three more small creeks to Tonahutu before it drains into Grand Lake.

A small, forked creek on the west side of Green Mountain flows into Onahu Creek before joining the Colorado River. The Green Mountain Trail follows this creek and then joins the Tonahutu Creek Trail on the east side of Green Mountain.

Columbine Lake, outside the Park, is fed by Harbison Ditch from Tonahutu Creek to the northwest of Grand Lake. Little Columbine Creek also flows from Columbine Lake to Shadow Mountain Lake.

Onahu Creek originates off a cirque at the divide between Mount Ida and Nakai Peak, with three branches, including waters draining out of Julian Lake. Onahu Creek flows to the southwest before joining the Colorado River in the Kawuneeche Valley. The Onahu Creek Trail comes off Highway 34 and follows this creek before joining the Timber Creek Trail from the north. Chickaree Lake is a small, isolated lake near the creek.

A small pond in Long Meadows joins Timber Creek to the north, as well as another creek draining Long Meadows to the south into Onahu Creek.

Timber Lake runs off the northwest flank of Mount Ida and is joined by several small creeks flowing off Jackstraw Mountain and north out of Long Meadows. Timber Creek drains into the Colorado River north of the Holzwarth Ranch adjacent to the Timber Creek Campground. The Timber Lake Trail comes from the north off Highway 34 and goes toward Timber Lake, as well as a trail junction heads south as the Timber Creek Trail goes through Long Meadows and joins the Onahu Creek Trail.

Beaver Creek drains off the northwest flank of Jackstraw Mountain and joins the Colorado River near the Timber Lake Trailhead off Highway 34 (Trail Ridge Road).

# Northwest Corner: Fall River Pass,
## Chambers Lake Quadrangles

The Colorado River, formerly the Grand River, has its headwaters on the west side of the Continental Divide in Rocky Mountain National Park. This is the dominant river of the west side of the Park.

It begins in the valley between Lulu Mountain to the west and Specimen Mountain to the east. The Colorado River drains from north to south from just south of La Poudre Pass to Shadow Mountain Lake and Lake Granby. The Colorado River winds its way south through the Kawuneeche Valley, providing lush riparian habitat with many ponds and wetlands along the way. It parallels Trail Ridge Road after it drops down from the pass, down the valley to the Grand Lake area.

Off the northeast slopes of Specimen Mountain, three creeks drain to become Willow Creek, which flows northeast to Long Draw Creek, outside the northwest corner of the Park.

Most of the following creeks flow off the eastern side of the Never Summer mountains, historically into the Colorado River before diversion, these are described from the north to the south.

Off Lulu Mountain, Bennett Creek, just south of La Poudre Pass was the first creek to feed the Colorado River from the west, and an unnamed creek and Specimen Creek drained into the start of the Colorado River Valley from the east.

Coming off the southeast spur of Lulu Mountain from the west is Lady Creek, draining toward the Colorado.

From the west, off Mount Richthofen, out of Box Canyon, flows Lulu Creek. The Thunder Pass trail comes off the divide, drops through Box Canyon, past the Ditch Camp and down toward the Colorado River.

Sawmill Creek comes from the west off Tepee and Lead Mountain, through Skeleton Gulch, south of the Ditch Camp toward the river.

Coming off Lead Mountain, Little Dutch Creek from the west and an unnamed creek from the east head toward the Colorado River near the historic site of Lulu City.

On the east side, from the Crater, between Specimen Mountain and Shipler Mountain, flow several unnamed streams out of Crater Gulch, toward the Colorado River.

From a high cirque on Cirrus Mountain, out of Lake of the Clouds, tumbles Big Dutch Creek, coming down Hitchens Gulch toward the Colorado River from the west.

An unnamed creek runs from the east off Shipler Mountain toward the Colorado, just north of Lost Creek.

Lost Creek flows off Howard Mountain on the west side, a small, isolated pond, Pinnacle Pond, sits on the southeast side of Howard Mountain. Mosquito Creek flows to the southeast from Howard, joining Opposition Creek, before draining historically into the Colorado River.

Opposition Creek drains off Mount Cumulus and Mount Nimbus, from the west, through Hells Hip Pocket.

Off the south spur of Specimen Mountain, near Milner Pass, Bighorn Lake drains into Squeak Creek, and into the Colorado River just north of Phantom Creek.

North of Trail Ridge Road (Highway 34), west of the divide, Lake Irene flows into Phantom Creek which drains to the southwest and into the Colorado River.

Beaver Creek streams from the east, coming off south of Trail Ridge Road and Farview Curve, off the northwest flank of Jackstraw Mountain, to the Colorado River near the Timber Creek Trailhead.

Red Gulch comes off the west between Mount Nimbus and Mount Stratus, historically flowing into the Colorado River between Beaver Creek and Timber Creek.

# Grand Lake, Bowen Mountain Quadrangles: west of Highway 34

South of Baker Mountain, off Parika Mountain, Parika Lake flows into Baker Gulch, before flowing toward the Colorado River. The Grand River Ditch's southern terminus includes the waters flowing out of Baker Gulch and those flowing off the Never Summer Mountain Range to the north. Fogelberg and Grinstead (2006) describe a nice hike to the Grand Ditch in the Baker Gulch area.[5]

The streams on the west side of the Park, beginning north to south, which were diverted to the Grand River (Grand) Ditch include: Specimen, Bennett, Lady, Lulu, Sawmill, Little Dutch, Middle Dutch, Big Dutch, Lost, Mosquito, Opposition, Red Gulch, Baker Creek, and Parika Lake. The Grand Ditch covers 14.3 miles, diverting these creeks from the Colorado (formerly Grand) River (Western Slope) to the Cache La Poudre River (Eastern Slope). This diversion is covered in detail in the chapter on Water Impoundments and Diversions in and adjacent to the Park.

# CHAPTER TWO

# Water Impoundments and Diversions in and Adjacent to the Park

The majority of early Colorado's population lived on the plains to the east of the mountains. Although the railroads had promised "the rain will follow the plow" in their land promotion for those coming to settle the west, this was not true. Stephen Long's "Great American Desert" description was more apt for the region on the map labeled

*Neníisótoyóú'u (Arapaho for The Two Guides), Long's and Meeker Peaks.*

this after his 1820 expedition. Jessie Fremont, the wife of explorer John C. Fremont, on June 30th, 1843, described the South Platte River near present day Sterling, as *"swollen with the waters of the melting snows"*. She further described, *"Long's Peak and the neighboring mountains stood out into the sky, grand and luminously white, covered to their bases with glittering snow"*. [1] This was the hope expressed for those traveling to this region. With an average of eight to fifteen inches of annual precipitation on the Front Range[2], the reality of civilization and agriculture was tied to those snow-covered peaks.

Most towns had sprung up along creeks and rivers. Camp Collins along the Cache la Poudre River near Laporte, relocated to become present day Fort Collins after a flood in 1864. Denver was established near Cherry Creek and also flooded in 1864. Boulder grew up around Boulder Creek, which flooded in 1874. The Union Colony (now Greeley), established by Nathan Meeker in 1870, was located near the junction of the Cache La Poudre (Poudre) and South Platte rivers.

Irrigation started here in 1844, when Antoine Janis settled on the banks of the Cache La Poudre River, followed by the Union Colony settlement in 1870. Union Colony began building extensive irrigation canals early on to bring water to their crops from the Poudre River. This type of irrigation system spread near present day Fort Collins and Longmont. These towns were all in the shadow of Long's and Meeker peaks, and potential downstream recipients of at least the waters flowing east of the Continental Divide. E.S. Nettleton, a Union Colony water expert, and other irrigation pioneers such as Benjamin Eaton, envisioned bringing more water from the mountains.[1]

In July of 1874, Union Colony irrigators discovered the Poudre River was dry at their Ditch Number Three headgate. After traveling 30 miles to Fort Collins, they discovered Fort Collins ditches had usurped their Poudre water. Described as an angry meeting over the disputed water at the Whitney schoolhouse, the Greeley group declared that they had "prior rights as against any Fort Collins canals". Fort Collins water users didn't want to yield any water or their rights to it, but finally relented to release some water downstream to Union Colony's headgate. This "collaboration over conflict" ended peacefully that night. Five days later, a rainstorm brought five inches of rain to the region, which temporarily solved the problem.[3]

This conflict has been described as "The Watershed Moment" that changed water law in the west. Two years later, in 1876, the Greeley group worked with Denver groups and other future Coloradans to shape the water provisions in the Colorado Constitution, which essentially became water law in the western United States. This Colorado Doctrine of Prior Appropriation outlined five principles of water law (see Glossary). Essentially water belonged to the public, but the right to beneficial use could be obtained, and where there wasn't sufficient water to satisfy all use rights, the earlier established right prevails (senior rights).[3] While this is oversimplified, this Right of Prior Appropriation has been the law of water in the west, with "water adjudication courts" to oversee it. This model was far different from "Riparian Water Law" in the east, where water is relatively abundant for
all users and needs and based on land ownership adjacent to the water.

In 1881, the newly organized Larimer County Ditch Company adopted Union Colony's irrigation model, and built the Larimer County Canal, bringing Poudre River water to thirteen-thousand acres northeast of Fort Collins.[4] The 1880s saw eastern U.S. and European investors bring irrigation from the South Platte River to the south and east, the Arkansas River to irrigate dry Rocky Ford, and Mormon colonies brought water to the San Luis Valley from the Rio Grande River.[1]

By the 1890's many men saw Colorado's mountains not for their scenic value, but for the water waiting to melt. It was a time of a more utilitarian view of the available resources of the mountains, rather than the preservation of the land for aesthetic purposes.[5]

One of the first modifications to hold or divert water near Rocky Mountain National Park was just north of the Park with the Chambers

Lake dam. Chambers Lake was a natural lake on Joe Wright Creek, that drains into the Poudre River. It was named for trapper Robert Chambers, who camped there with his son in 1858. Robert Sr. was killed by Native Americans at their camp, while his son was away getting ammunition and supplies. Robert Jr. later told the Union Pacific Railroad about the timber at this location for use as railroad ties in 1867. It was the timber workers camped there for logging that named the lake in honor of Robert Sr.[6] Timber cut for railroad ties was run down the Poudre River from Joe Wright Creek through Poudre Falls to Chambers Lake to wait for spring runoff to send the logs the rest of the way down river. Ties were also cut from the Little South Fork of the Poudre near present day Pingree Park. The tie industry was ended in the early 1880's when the bulk of the eight-inch timber had been cut.[7] Because of a lack of good roads, the river was the transport means for the ties to get to LaPorte, where they were offloaded by a boom for the Union Pacific Railroad. Chambers Lake was later leased by the Larimer County Reservoir Company to the ditch company. The Larimer County Ditch Company started increasing the natural lake's capacity in 1887, by deepening the lake and constructing a ten-foot earthen dam. The dam broke on June 8 1891, taking out every bridge on the Poudre River, and destroyed Poudre City. There were no human fatalities. In the litigation that followed, the ditch company dissolved and was purchased by the Water Supply and Storage Company in 1892.

On May 20, 1904, the Chambers Lake dam broke again, and resulted in at least one death, George Robert Strauss. George Strauss was one of the founding members of the Fort Collins area in 1860. His homestead and cabin were near present day Timnath.[8]

In the spring of 1907 the dam broke again, with no loss of life, and all flood claims were settled without the courts and without serious financial losses to Water Supply and Storage Company (WSSC).

Water Supply and Storage rebuilt the dam in 1910, and in 1922 again started rebuilding to enlarge the reservoir capacity. The project was to eventually raise the dam height to 58 feet. This was completed in

1928. In 1974, the State Engineer's Office imposed a storage limit on the reservoir, limiting the gage height to 45 feet. In 1983, Chambers Lake Dam was classified by the State Engineer's Office as a High Hazard Dam.[9]

## Skyline Ditch

The Skyline ditch was under construction to divert water from the Laramie River to Chambers Lake between 1891 and 1895. In 1894 the first water was diverted through Skyline from the Laramie River, which normally flows into the North Platte River in Wyoming, to Chambers Lake, and the Cache La Poudre River, which empties into the South Platte River. The final Skyline Ditch length was five miles. Norman Walter Fry, in his book, details the construction of this ditch, which came under the ownership of the Water Supply and Storage Company.[7]

## Laramie-Poudre Tunnel

The second diversion of water from the Laramie River to the Poudre was the Laramie-Poudre Tunnel project, built between 1909 and 1911. The tunnel length is two miles and work involved crews working from nearby camps with power supplied by three water wheels at Poudre Falls to drive drills and provide lighting. Working from both ends of the project, the tunnel bores met with less than an inch of error.[6] WSSC and the Windsor Reservoir and Canal company formed the Tunnel Water Company to manage the Tunnel and facilities.

## Rawah Ditch

An Upper Rawah Ditch and West Side Ditch divert water from tributaries on the western side of the Laramie River, and the East Side Ditch carries water from eastern tributaries of the river. Collectively the Rawah Ditch gathers waters from streams emptying into the Laramie River and brings it north to a diversion structure at the Laramie- Poudre Tunnel that then empties into the Poudre River.[9]

## Michigan Ditch

The Michigan Ditch is north of the Park boundary at Cameron Pass. This ditch captures water from Lake Agnes (Island Lake), and takes it seven miles to Cameron Pass, where it empties into Joe Wright Creek. It picks up the water from the tributaries of the Michigan River, which normally would flow into the Laramie River and then to the North Platte River in Wyoming.[6] This is still eastern slope water but diverts to the Poudre River and ultimately the South Platte River, which eventually meets the North Platte in Nebraska to form the Platte River.

Originally the Skyline Ditch could deliver up to 18,000 acre-feet of water to the Poudre River, but litigation with the State of Wyoming reduced all diversions from the Laramie River to a total of 19,875 acre-feet. These included the Laramie-Poudre tunnel, Skyline and Michigan ditches.[9]

*Water flowing into the Specimen Ditch. Photograph courtesy of Dave'n'Kathy's Vagabond Blog 2017.[10]*

## Grand River Ditch

In 1894, the Water Supply and Storage Company (WSSC) agreed to begin construction of five miles of the Grand River Ditch to bring water that flows into the Grand River (Colorado River) into a ditch that would divert these waters to the east side of the Continental Divide.[4] Of the ditches described in this chapter, the Grand River Ditch, Specimen Ditch, and Harbison are the only ditches that actually divert water directly out of Rocky Mountain National Park.

This project was started in 1891 by the Larimer County Ditch Company, and plans were to divert creeks coming off the Never Summer Mountains that flow into the Grand River, which drains on the west side of the Continental Divide. The diversion would bring west slope water across a low pass at La Poudre Pass to the Poudre River flowing east.

Specimen Ditch was dug on the northwestern flank of Specimen Mountain in 1898 from Specimen Creek to La Poudre Pass, bringing the creek water to the northeast instead of it draining into the Grand (Colorado) River. The first water was delivered from this ditch in 1900.

This was one part of the larger project that became the Grand River Ditch. It joins the current Grand Ditch just before the pumphouse at La Poudre Pass Ranger Station.

*Grand Ditch and La Poudre Pass Ranger Station. Photograph courtesy of Dave'n'Kathy's Vagabond Blog 2017.[10]*

The first section of the Grand River Ditch completed was the Bennett Ditch. In 1896, WSSC cleared timber from the north and south sides of the ditch path. In 1898, work on the south side ditch over Poudre Pass to the Poudre River was started and completed in 1900, this ditch was 1.22 miles long. In September 1900, work was started on the north side ditch with hope that work would be completed in 1901. This ditch was to be started at Dutch Town Creek and would run two and one-half miles to Sawmill Creek, and one-half mile to Lulu Creek. Only one mile was completed in 1901 due to the short working season. With money problems and a contractor dispute, delays to construction occurred until more funding and insurance was secured.[4]

Most of the early ditch work was completed by hand, with laborers using picks and shovels, wheelbarrows, and blasting rock with black powder. Teams of mules and horses also pulled scrapers to dig the early portions of the ditch.[11]

The WSSC began granting small contracts to complete sections of the ditch in 1903. By the end of 1904, the ditch was complete one-half mile past Dutch Town Creek. Repairs to the Larimer County Canal took precedence following the flood of 1904. During 1906 the ditch was completed to Tank Creek. Construction was then delayed on the ditch for the next few years.

In 1906, the State Engineer inspected the work on the Grand River Ditch and asked for detailed construction plans for the planned Reservoir on the Cache La Poudre (Long Draw Reservoir). La Poudre Creek was the creek that became Long Draw Creek.

In 1907, 192,000 feet of timber was removed from the path of the proposed extension of the ditch. No actual ditch construction occurred until after 1909, partially because of the Chambers Lake Dam 1907 breach.

In 1911, a contract was let for the rebuilding of the Chambers Lake Reservoir. In 1913, the channel for Chambers Lake was cut and cribbing was done on the Grand River Ditch. WSSC secured lumber from the Forest service in 1914 to cover the ditch.

In 1915, a guarantee fund was established by Fort Collins and area to build a road up the Poudre Canyon, $100 was contributed by WSSC, since that would provide better access to their mountain projects. Rocky Mountain National Park was also established in 1915, and the U.S. Congress was contemplating increasing the Park size, which WSSC objected to, since it would interfere with the possible extension of the company's water rights.[3] When Congress finally acted in 1924, the rights of WSSC were protected. During the establishment of the Park, negotiations had been ongoing to make sure the boundary of the Park excluded areas deemed critical to ongoing water projects. Establishment of Long Draw Reservoir was part of the north boundary change of Rocky Mountain National Park.

In 1917, 9,000 feet of the Grand River Ditch had been covered. The covering of the ditch was to prevent mountain slides from blocking the ditches. During World War I, labor was in short supply and the WSSC only made repairs to their ditch system that were necessary.[4]

In 1919, with an "extreme drouth in the Cache La Poudre Valley", the company decided that obtaining more water from the Grand River Ditch was worth the investment of $25,000 per mile to extend the ditch to Baker's Gulch. No action was taken until 1921, and a saw mill was purchased and put up next to the Grand River Ditch to save money.[4]

Nineteen-twenty-one also was the year that the Grand River was renamed the Colorado River officially by Congress. Eventually the Grand River Ditch became the Grand Ditch.

No more work on extending the ditch occurred until 1928, when work on the ditch continued, as well as construction on the Long Draw Reservoir. Long Draw reservoir had a 58-foot dam and had a capacity of 5,700 acre-feet when it was completed in 1930.

The Grand Ditch was extended in 1933-1934 to Lost Creek. By fall of 1936, the last six miles was completed to Baker Gulch using steam shovels. The ditch from initial construction to completion took almost 47 years. It measures 14.3 miles long, is twenty feet wide, and six feet deep. It has a carrying capacity of 360 cubic feet per second. It delivers approximately 20,000 acre-feet of water a year to the Front Range.

Over the years, besides ditch bank breaches, water leakage through porous surfaces resulted in the WSSC attempting to seal the ditch with clay hauled in with small trucks, and placing a "liner membrane" in 1974, to help keep the water in the ditch with minimal loss prior to reaching the storage reservoirs.[9]

# Summary of the Grand Ditch

In 1976, the historic Grand River Ditch (Grand Ditch) and Specimen Ditch were nominated in the National Register of Historic Places.

In the description for the nomination, the ditch was summarized as follows[11]:

> *The Grand Ditch is sited on the precipitous eastern flank of the Never Summer Range. It starts at Baker Creek (elevation 10,300 ft.), runs northeastward through a rocky terrain covered intermittently with a spruce-fir forest, gathers water from Baker Creek, Red Gulch, Opposition Creek, Mosquito Creek, Lost Creek, Big Dutch Creek, Middle Dutch Creek, Little Dutch Creek, Sawmill Creek, Lulu Creek, Lady Creek and Bennett Creek, and discharges into [a] La Poudre Pass Creek at La Poudre Pass (elevation 10, 175 ft.).*
>
> *Specimen ditch runs from Specimen Creek (elevation 10,300 ft.) to La Poudre Pass, passing through a spruce-fir forest on the northwestern flank of Specimen Mountain.*
>
> *The Grand Ditch is an earthen canal approximately 14.3 miles long. The cross section of the ditch widens, obviously, as one approaches La Poudre Pass; the cross section is trapezoidal, about 20 feet wide and 6' feet deep. An unimproved road, running along the berm, parallels the Grand Ditch for mainte-nance access. The Specimen Ditch, somewhat smaller than the*

*Grand Ditch because of the smaller volume of water, is approx-
imately 1.7 miles long.*

*Also included in this nomination is Camp 2, a work camp
built about 1989 to house the workmen who constructed this
segment of the Grand Ditch. Camp 2 is located about 400' west
of the ditch in a marshy meadow, and surrounded by a spruce-
fir forest. Nine cabin ruins, of saddle notched and V notched log
construction, were found at this site. A broken range identifies
one of the cabins as a cook shack, and pieces of slag mark the
blacksmith shop."*

Often misidentified in the historical literature were the "Chinese
Coolies" working on the ditch. My coworkers at WSSC in 1974 told
me the workers were Chinese, and that the buried workers graves were
desecrated, having had their teeth knocked out to pilfer the gold fillings.
Water Supply records in 1904 reference Japanese workers hired in "companies". *The Register*[11] further describes,

*"One of the most interesting features of Camp 2 is the series
of Japanese "dugouts" in the hillside south of the meadow. It
is said that the Japanese, unaccustomed to American food,
drew food supplies and cooked for themselves, separate from
the other workers. These dugouts were, supposedly, used by the
Japanese for their cooking (and food storage-Author's note),
and adjacent "ovens" may have been used for the making of
charcoal."*

*"The Grand and Specimen Ditch, owned by the Water
Supply and Storage Company of Fort Collins, will continue
to be used for water diversion and maintenance will be per-
formed. The National Park Service will continue to maintain
bridges, trails, signs, a ranger station, and a shelter cabin in
and adjacent to the nominated area".*[11]

John McNabb was the chief foreman for much of the early Grand Ditch construction and was also instrumental in the construction of the Skyline Ditch, Michigan Ditch, Chambers Lake Dam, Joe Wright Reservoir, the Laramie-Poudre Tunnel diversions, as well as repairs to bridges and the Poudre Road. Norman Walter Fry describes him as *"the "indispensable man", "and probably did more "on the job" work to develop the water resources of the Upper Poudre than any other man."* Fry also described McNabb *"as the finest axe man we ever had on the River. It was a pleasure to see the way he could roll up a log cabin while other men were thinking about it".*[7]

Every year the WSSC would plow Long Draw Road and the ditch road to reach the ditch before spring breakup. The goal was to make sure the ditch did not have ice dams that impeded the flow, and cause water to overflow and wash out the ditch.

In the spring of 1974, I worked for WSSC as a ditch laborer at $2.25 per hour. Two other men and I went up that year to plow Long Draw Road to the ditch, I was the cook, fuel truck and trailer driver, and the others plowed the road using D6 and D7 Caterpillar tractors. Plowing progress was slow that year, and when we finally reached the

*Plowing into the Grand River Ditch, April 1974.*

ditch, it was already washing out. We drove to town to get the "Drott" (bucket scoop) to clean out the ditch. That day, when I received my paycheck, the per diem charges for food while camped out took most of my paycheck, so I quit.

About a month later I went to work for Rocky Mountain National Park. While working there that summer, I was told by the RMNP Road Crew folks that the Park Service ended up helping with much of the repairs to the ditch and access road that year. I don't know how many other ditch breaches have occurred in the last 83 years since completion of the ditch, but a large breach brought the attention of the National Park Service under the Park System Resource Protection Act.

On May 30, 2003, a large ditch breach occurred 2.4 miles south of La Poudre Pass. It saturated the hillside and gave way, sending 48,000 cubic feet of mud and rocks down the hillside into Lulu Creek and into the Colorado River.

According to the National Park service approximately 22 acres and 1.5 miles of stream and habitat were injured. The Department of Justice, representing the NPS filed a civil lawsuit against the WSSC. In May of 2008, an out of court settlement was reached, where WSSC agreed to pay RMNP nine million dollars in damages. RMNP began a multi-year Environmental Impact Assessment to guide the restoration of the breach area. Public scoping meetings were scheduled in June 2010.[11]

*2003 Ditch Breach, Photograph National Park Service.*

During the original construction of the Grand River Ditch, some of which occurred prior to the designation of Rocky Mountain National Park in 1915, the ditch created a very visible scar across the Never Summer Mountains. The scar has increased in visibility where minor ditch breaches have occurred over the years.

*Grand Ditch scar across the Never Summer Range-2019. Look especially at scar on middle right of picture.*

Ben Fogelberg and Steve Grinstead,[13] wrote about hiking the Grand Ditch in *Walking into America's Past, 50 Front Range History Hikes. Hike 6, Baker Gulch to The Grand Ditch,* has a nice description of the ditch, its history and outlines a nice hike up Baker Gulch to the ditch at the southern end of the project. The northern end of the ditch and Specimen Ditch can be hiked after driving up Colorado Highway 14 (Poudre Canyon) to Long Draw Road to access the ditch road at the National Park boundary near La Poudre Pass Ranger Station.

## Harbison Ditch

Columbine Lake, to the northwest of Grand Lake, outside the Park, is fed by Harbison Ditch from Tonahutu Creek. This diversion was part of the Harbison Ranch, homesteaded in the 1880's. This has been purchased by the National Park Service.

## Eureka Ditch

In 1902, the Eureka Ditch was dug on the top of the continental divide at Bighorn Flats, between Ptarmigan Point and Sprague Pass. This ditch diverted western-draining water to Spruce Canyon to the east. This is one of the eighteen projects pre-dating the Park that Superintendent Roger Toll wanted to mitigate in 1923.[5]

# Colorado-Big Thompson Project and Alva B. Adams Tunnel

Park Historian C. W. Buchholtz details an ambitious plan to tunnel under the mountains from Grand Lake to Moraine Park, put forth by engineering students at Colorado State College (now Colorado State University) in 1905.[5]

In that same era, the U.S Reclamation Service withdrew land around Grand Lake for future projects and suggested raising the elevation of Grand Lake by twenty feet. This plan would include a reservoir to store 140,000 acre-feet of water and construct a 12-mile tunnel from Grand Lake to either the Big Thompson or Saint Vrain Rivers.[14]

When Rocky Mountain National Park was created in 1915, the Reclamation Service was assured that right of ways for water diversions and other projects would not be "disrupted" by the new Park.[5]

In 1922, the federal Colorado River Compact apportioned the Colorado River's waters between seven upper and lower river basin states. In the 1920's, the Dust Bowl withered crops throughout eastern Colorado and the west and furthered the pressure on Congress to approve a trans-mountain diversion project. Colorado's Democratic Senator, Alva B. Adams, was the greatest proponent of a *"mammoth water project"* that *"would benefit not only the state's agri-business, but its other local economies as well"*.[14]

In 1935 the Public Works Administration approved an allotment of $150,000 to the Bureau of Reclamation (renamed 1923) to survey the

Grand Lake-Big Thompson Project. The National Park Service strongly disagreed with allowing a tunnel in the Park, citing a list of potential damage to the Park, harm to wildlife, and worries of decreased visitation, among other problems. The need for preservation was cited to protect Rocky and other parks from suffering from the scars of civilization.[5]

A compromise was made to ensure western Colorado received acre-foot for acre-foot compensation for waters taken from the Colorado River. This resulted in a 152,000 acre-feet compensatory storage reservoir named Green Mountain Reservoir. This would be located 13 miles from the town of Kremmling on the western slope. The Bureau of Reclamation renamed the project to drop the name "Grand Lake" from its title, and the project became the Colorado-Big Thompson Project (C-BT). This served to decrease public antagonism "to a certain class of people" opposed to the project.[14]

In 1937, with congressional approval of the funding for C-BT in the Interior Department Appropriations Act, final testimony on November 12, 1937 by Interior Secretary Harold Ickes was headlined by the Denver Post as: "Ickes Says he is Forced to Favor It".[14]

Construction was contingent on creating Colorado's first water district, The Northern Colorado Water Conservancy District (NCWCD). This allowed for holding property, levying taxes and assessments, allocation of the C-BT water, and contracting with the Federal Government.[14] In a final approval by President Roosevelt in December 1937, Secretary Ickes pledged "under my direction as Secretary of the Interior the interests of those devoted to the cause of our national parks will be protected".[14]

The Green Mountain Dam was built between 1938 and 1943. It had a capacity of 154,645 acre-feet of water and produced 25.8 megawatts of power when completed.

The "Alva B. Adams" Tunnel began construction on June 16, 1940, from both the east side and the west side. In late 1942, work was halted by the War Board, as non-essential to the war effort. The tunnel work began again in September 1943, and the two ends of the tunnel met on

Table 1: **Dams of the Colorado-Big Thompson Project**

| Dam | Hydraulic Dam Height (ft) | Crest Length (ft) |
|---|---|---|
| Green Mountain | 264 | 1150 |
| Granby | 223 | 861 |
| Willow Creek | 95 | 1100 |
| Shadow Mountain | 37 | 3077 |
| Mary's Lake | 20 | 820 |
| Olympus | 45 | 1951 |
| Rattlesnake | 100 | 1100 |
| Flatiron | 55 | 1725 |
| Carter Lake | 190 | 1235 |
| Horsetooth | 111 | 1840 |
| Soldier Canyon | 203 | 1483 |
| Dixon Canyon | 215 | 1265 |
| Spring Canyon | 198 | 1120 |

**Reservoirs of the Colorado-Big Thompson Project**

| Reservoirs | Reservoir Capacity (ac-ft) | Miles of Shoreline |
|---|---|---|
| Green Mtn. (west) | 154,600 | 19 |
| Lake Granby (west) | 539,800 | 40 |
| Willow Creek (west) | 10,600 | 7 |
| Shadow Mtn (west) | 18,400 | 8 |
| Mary's Lake (east) | 900 | 1 |
| Lake Estes (east) | 3,100 | 4 |
| Pinewood (east) | 2,180 | 3 |
| Flatiron (east) | 760 | 2 |
| Carter Lake (east) | 112,200 | 8 |
| Horsetooth (east) | 151,800 | 25 |

*Modified from Table 1: Autobee, R. 1996. Colorado-Big Thompson Project, p 23. Source:* U.S.D.I., B.O.R, Colorado- Big Thompson Project Technical Record of Design Construction, Vol. 2. *Washington, D.C., Government printing Office 1957, vi.*

July 10, 1944. Many writers have commended the project for both ends meeting with close precision, but my grandfather, a Bureau of Reclamation engineer, would have expected no less. The nine-foot, nine-inch tunnel was reinforced and concrete- lined one-foot thick.

Finally on June 23, 1947, the first water of the C-BT project flowed from Grand Lake to the East Portal near the YMCA Camp in Estes Park. A 69,000-kilovolt transmission line, encased in a pipe on the roof of the tunnel provided power between the east and west slope power facilities.[14] This "submarine cable" line prevented the need for a "giant mountain climbing transmission line" up and over the Continental Divide in the Park.[14] The 13.1-mile tunnel itself is the only part of the C-BT project in Rocky Mountain National Park itself. The West Portal and the East Portal are outside of the Park boundaries.

The associated dams, dikes, and pumping structures for the diversion project were built between 1941 and 1956. Besides the powerplant at Green Mountain, five more powerplants were built in the project.

The flow of water in the project goes as follows. Lake Granby is ten miles downstream from Grand Lake. Granby Dam and four dikes collect water from the Colorado River drainage and retain it for pumping into Shadow Mountain Reservoir and Grand Lake. Additional water for Lake Granby storage comes from Willow Creek, a Colorado River tributary below Granby Dam. Willow Creek Reservoir collects this water, and through a feeder canal two miles long, is lifted by pumps 175 feet into Lake Granby. Lake Granby has a 12-story high pumping station, partially submerged, in the northeast corner of the reservoir. Shadow Mountain Reservoir is between Grand Lake and Lake Granby and fills with water from the north fork of the Colorado River. Lake Granby water is pumped into Shadow Mountain through the pumping station via a canal, 1.8 miles long. From Shadow Mountain Reservoir the water flows into Grand Lake, and then over a fixed weir into the mouth of the west portal of the Adams Tunnel. This is the western component. One intent of this project was to avoid fluctuations in water levels in Grand Lake, which is the only natural lake in this project.

On the east side, water comes out of the tunnel at the East Portal, 750 feet downhill from the West Portal, into a pond dammed by the East Portal Dam. This is pumped 1.3 miles by the Aspen Creek Siphon into the Rams Horn Tunnel, then through a short penstock into Mary's Lake. Mary's Lake has Dikes 1 and 2 and the reservoir as an afterbay, for the Mary's Lake Powerplant. From Mary's Lake, water travels six-tenths of a mile through a conduit to the Prospect Mountain Tunnel. This tunnel provides power through underground pipes to the Estes Powerplant. Water then exits into Lake Estes, where some is diverted to the Big Thompson River down the canyon, and some water is diverted through a series of tunnels, to generate electricity at Pole Hill, Flatiron, and Big Thompson power-plants. From Flatiron Reservoir, water travels north to Horsetooth Reservoir and the Poudre River through the Charles Hansen Feeder Canal (FC), formerly Horsetooth FC. The canal crosses the Big Thompson Canyon 1.5 miles upstream of the Big Thompson (BT) river in a nine-foot-diameter steel siphon pipe. This siphon provides 930 cfs of water to the BT River, and 550 cfs to Horsetooth Reservoir[13]. While these tunnels, penstocks and dams for powerplant afterbays need a great deal of power to operate pumps, the system also provides 750 million kilowatts of electricity annually, enough to power over 58,000 homes a year. The Colorado-Big Thompson Project in its first year of operation (1957) serviced 720,000 irrigated acres in northeastern Colorado.[15]

An expansion to the C-BT Project, called the Windy Gap Project began construction in 1981, and was completed in 1985. This was a project designed to provide municipal water to six participants, the cities of Boulder, Estes Park, Fort Collins, Greeley, Longmont, and Loveland. Windy Gap is a diversion dam on the Colorado River with a 445-acre-foot reservoir, a pumping plant and a six-mile pipeline to Lake Granby. The stored water at Windy Gap Reservoir is pumped to Lake Grandby, before being delivered to the Alva B. Adams tunnel and then to the eastern slope Colorado-Big Thompson Project.

An additional part of this project is the proposed Chimney Hollow Reservoir, west of Carter Lake in Larimer County, that will store 90,000 acre-feet of capacity to supply 30,000 acre-feet of water each year for twelve northeastern Colorado water providers. The intent of this project is essentially just reliable storage for already allocated Windy Gap water. Ground-breaking for this reservoir was done on August 6, 2021. This project is also called the Windy Gap Firming Project and is under the direction of the Northern Water Municipal Subdistrict.[16]

A modification to the Windy Gap Project is beginning in Fall of 2021, where the emptied Windy Gap Reservoir Dam is being modified to relocate a section of dam from the south to the north. This will be done in order to provide a better channel between the northern (upper) portion of the Colorado River to the southern (lower) section of the Colorado River. This will provide an enhanced fishery, as well as better wildlife habitat for the area. The project timeline is starting in spring of 2022 with completion planned for 2023.

The Colorado-Big Thompson Project does tunnel directly through Rocky Mountain National Park and brings western slope water to farmers and water users on the eastern plains, using Colorado River water that originates in the Park. However, the greatest benefit to the Park may be the added recreation areas that the reservoirs in the C-BT Project provide. This recreation use potentially decreases the impact of visitors to the Park, by creating alternatives to visiting the National Park. As our Colorado population has exploded since the C-BT project was originated, the pressure on outdoor recreation has grown exponentially with that population.

# CHAPTER THREE

# Dam Structures in
# Rocky Mountain National Park

There were a large number of water diversions and reservoirs that had been approved prior to the founding of Rocky Mountain National Park in 1915. While many of these had not yet been developed or completed by the time the Park was dedicated, the enacting legislation creating RMNP protected the existing water rights under the Act of 1891, which provided a five-year deadline for completion of the water projects.[1]

WATER DIVERSION AND RESERVOIRS APPROVED BEFORE THE FOUNDING OF ROCKY MOUNTAIN
NATIONAL PARK (1915 BOUNDARY)

| Date Approved | System | Purpose |
|---|---|---|
| August 20, 1903 | Eureka Ditch | Water Diversion |
| November 3, 1903 | Milner Pass Ditch | Water Diversion |
| December 19, 1903 | Lawn Reservoir | Irrigation |
| February 26, 1904 | Ouzel Reservoir (Arbuckle #1) | Irrigation |
| February 26, 1904 | Pupit Reservoir (Arbuckle #3) | Irrigation |
| February 26, 1904 | Bluebird Reservoir (Arbuckle #2) | Irrigation |
| June 10, 1905 | Virginia Reservoir | Power |
| January 29, 1906 | Cairns Reservoir No. 2 | Power |
| January 29, 1906 | Cairns Reservoir No. 3 | Power |
| December 24, 1907 | Cairns Reservoir No. 1 | Power |
| October 1, 1910 | Sandbeach Reservoir | Irrigation |
| May 31, 1911 | Deer Mountain Reservoir | Irrigation/Fishing |
| September 26, 1913 | Glacier Reservoir No. 1 | Irrigation |
| September 26, 1913 | Glacier Reservoir No. 2 | Irrigation |
| March 16, 1914 | Green (Thunder) Lake Reservoir | Irrigation |
| March 16, 1914 | Green (Snowbank) Lake Reservoir | Irrigation |
| March 16, 1914 | Long (Eagle) Lake Reservoir | Irrigation |
| March 16, 1914 | Box Lake Reservoir | Irrigation |
| July 2, 1914 | Pear Lake Reservoir (Arbuckle #4) | Irrigation |
| July 2, 1914 | Hutchinson Reservoir #1 | Irrigation |
| July 2, 1914 | Hutchinson Reservoir #2 | Irrigation |

Enacting legislation creating Rocky Mountain National Park protected the existing
water rights under the Act of 1891 which also proscribed a five year deadline for
completion of the water projects. [5]

*Table 1. Pre-Rocky Mountain National Park Water Projects-table from page 7, National Register of Historic Places, National Park Service. 1984. E.O.11593. Determination of Eligibility Notification. Bluebird Dam.*

As a result, many of the projects listed on Table 1 did not get developed or completed.[2]

Superintendent Roger Toll inspected these projects in 1923, and after forwarding the inspection report to the General Land Office, the reply was "This office is now taking up with a view of rapidly disposing of all the uncompleted projects…giving status of various grants in Rocky Mountain National Park".[1]

## Bluebird Lake (Arbuckle Reservoir #2)

In 1902, Emma Arbuckle, and her son Frank, filed claims on five lakes in Wild Basin, including what were later named Bluebird, Pear, Sandbeach, and Snowbank Lakes. They owned a ranch on the North Saint Vrain River, west of Lyons, and were trying to secure supplemental water for dry years. They began to build dams on Pear and Bluebird Lakes in 1902. A few years later they sold the rights to a group of businessmen from Longmont, who formed the Arbuckle Reservoir Company. Bluebird Lake was originally named Pipit by William Cooper in 1908, but the name was changed to Bluebird by the U.S. Geological survey on their 1915 map.[3] The Arbuckle Reservoirs are located in Wild Basin, south of Long's Peak, in the southeast corner of what became Rocky Mountain National Park in 1915.

The Arbuckle Reservoir Company built a larger dam on Bluebird Lake, between 1914 and 1919. They measured the pre-dam depth of the lake at 50 feet. They hauled in 2000 bags of concrete and one-inch diameter steel rebar, but with a lack of sand to mix with the cement, had to haul in a disassembled granite crusher and an automobile engine to power it.[3] The final dam height was 58 feet, and dam length was 200 feet, completed in 1923. The maximum capacity of the reservoir with the concrete dam was 991 acre-feet (ac-ft).[4] A 46.5-foot-wide depression, one foot deep, in the crest of the gravity dam section, formed the spill-way. There were two outlets, the lower outlet was approximately 56 feet below the crest of the dam, and the upper-level outlet was centered on the dam, and approximately 25 feet below the crest. The lower outlet was approximately six feet below the original stream channel.[1]

Bluebird Lake is greater than six miles in the backcountry and is a long hike and was a difficult trek for horses and mules that hauled in the materials for the dam. The mules carried two bags of cement each, and the rebar bundles were balanced on an axle behind four horses or mules, with the end of the bundles dragging behind on the ground.[1,3] Part of the old road used to haul materials can be found below the Ouzel Lake-Thunder Lake Trail junction, which heads straight down the hillside to North Saint Vrain Creek. Following the fisherman's trail from Pine Ridge can take you to that route, but the crossing is difficult to find, unless you come straight down the hill from the trail junction. In the 1970's, my trail crew found what appeared to be part of the old log bridge structure across North Vrain Creek, for that road.

*Bluebird Reservoir Dam. Photograph- Rocky Mountain National Park.*

In 1933, the Arbuckle Reservoir Company sold the Bluebird water rights to the City of Longmont, along with the rights to Pear and Sandbeach Reservoirs for $92,000.[3]

I remember in the 1970's, a rock and metal shed built into a granite cliff-face, this was used by the work crews during original construction.

The granite crusher that I saw, I thought was below the Pear Reservoir dam. In 1974, I also thought that the Bluebird dam had a big gaping hole in it, that kept water from filling the dam anywhere close to capacity. What I likely saw was the lower outlet draining, which would flow continuously into the creek, if the valve was open, as long as there was water in the lake. It was reported to be inoperable in 1974 by the State of Colorado, with the extent of valve opening unverified.[1]

## Pear Lake (Arbuckle Reservoir #4)

The Arbuckle's began construction of a dam in 1902 at Pear Lake. Joe Mills had named this lake for its shape. In 1906, an amended filing for Arbuckle Reservoir #4, "*specified a lake depth of 15 feet, and a dam height of 11 feet, for an estimated cost of $400.*"[3]

In 1933, it was purchased by the City of Longmont, along with Bluebird and Sandbeach Reservoirs, for $92,000.[3]

In 1974, when I first saw the reservoir, water was up to the dam. In that time frame, a ranger accidentally dropped his radio into the water from the dam, and several rangers SCUBA dived to retrieve it. By some accounts, the dam had leaked enough that the original Pear shape of the lake was visible.[3] I also remember a large granite crusher sitting adjacent to the dam.

The Dam Safety Program of the National Park Service listed the Pear Lake Dam as a rockfill embankment, with a height of 28 feet and a maximum capacity of 451 ac-ft.[4]

## Sandbeach Lake (Arbuckle Reservoir # 1)

Sandbeach Lake was named by Enos Mills for the white sandy beaches surrounding it. Sandbeach Lake was one of the Arbuckle Lakes that claims were filed on in 1902. When the dam was built, most of the beaches were covered up. The road into the lake had "deteriorated into a trail" by 1916.[3] This road I believe came in from the Meeker Park area and followed along the top of the Copeland Moraine, avoiding the steep switchbacks of the current trail from the Wild Basin valley.

It was included in the sale of the three reservoirs in Wild Basin to the City of Longmont in 1933. The dam was built of rubble rock/masonry, and I remember masonry/rock structures that must have been outlet valve housings on the dam end of the reservoir in the 1970's. When I camped there with my dad in the mid-1970's, there was a beaver swimming in the lake, and there were so many commercial jets flying over on approach to Stapleton (Denver) Airport, that we re-named the lake "Sandbeach International".

Baker and McCormick listed the dam as being 25 feet high, with 369 ac-ft maximum capacity.[4]

## Snowbank Lake (Arbuckle Reservoir #5?)

This small alpine lake above Lion Lake # 2 in Wild Basin has little information available. It was described as having a very shallow dam, which would present no danger if it would break, according to Park Superintendent Chester Brooks (1983). I can find no listing for capacity or dam size and height. I don't recall seeing a dam when I was there, but it may have just been a natural appearing dam utilizing local rocks.

## Thunder Lake (Head Gate)

Although Thunder Lake in Wild Basin appears to be an undammed natural lake, there was a head gate installed in the early 20th Century (unknown date) to control the flow of water to the North Saint Vrain Creek. A listing of Water Diversions and Reservoirs approved before the founding of the Park lists Thunder Lake as Green Lake Reservoir, approved for Irrigation on March 16, 1914.[1] This headgate was removed after the Lawn Lake Flood.[5]

## Copeland Lake (Reservoir)-Wild Basin Area

Copeland Lake was a small natural, spring-fed lake in the valley just off of Copeland Moraine, where the current entry to Wild Basin is located.

John B. Copeland homesteaded 320 acres near this lake in 1889. His homestead was "proved up" in 1896. Copeland dug a ditch from the North St. Vrain creek to the lake and enlarged the pond with an earthen dam[3]. A lodge was built nearby, prior to the establishment of the National Park, the ownership shows multiple owners including Copeland between 1910 and 1914.

*The head gate at Thunder Lake was used to control the flow of water released into North St. Vrain Creek. Photograph, Rocky Mountain National Park.*

In 1913 the City of Longmont bought 129 acres around Copeland Lake and filed to build a reservoir. The dam costs were estimated at $63,000, and the dam was to be 43 feet high. Longmont secured the water rights and still uses the water today. The dam washed out in 1934, and was replaced by another earthen dam.[3,5] In 1949, my father worked in Allenspark, and said that ice blocks cut from Copeland Lake in winter, were stored in sheds for use in ice boxes for refrigeration in the local area.

For years the lake was used as a picnic ground and campground on the road to the Wild Basin area. In 1982, in a NPS internal report dated just three weeks before the Lawn Lake failure, Sandbeach Lake was reported as leaking, and presented sufficient hazard that the Sheriff's Office evacuated the Copeland Lake Campground.[4]

## High Drive Water Supply

Pieter Hondius moved to Estes Park and ran a cattle operation in Upper Beaver Meadows on 2000 acres beginning in 1900. He subdivided the land on High Drive and built a water diversion structure on Beaver Brook with a settling tank and water filter.

He built a pipeline 2.75 miles long on the south side of Deer Mountain that emptied into a small Reservoir on Buck Creek. This supplied water to the High Drive subdivision. The National Park Service purchased the water system and homes on High Drive in 1966 that were in the Park Boundary. The homes were then placed on the Estes Park water supply system.[5]

## Hallowell Park Pond
## (renamed Hollowell Park, 1988)

From 1933-1942, the Civilian Conservation Corps were camped in various locations in the Park, including Hallowell Park, where a water retention pond was dammed on Mill Creek, likely as a fishing pond for these men. It was later removed as well as many other structures of this era, leaving only the concrete overflow structure.[5,6]

*Hondius Water Diversion Structure. Photo: Kenneth Jessen, Loveland Reporter Herald*

## Endovalley Pond

There was a fishing pond with an earthen dam east of the campground (now picnic ground), that was removed, likely in the 1960's.[5]

## Moraine Park Reservoirs

Part of the Stead's Ranch (formerly Sprague) buildings included two small reservoirs above Steads, near the switchbacks on the Bear Lake Road. These were removed along with the majority of buildings in the 1960's as part of "Mission 66".[5,6]

## Lawn Lake (Reservoir)

Lawn Lake occupies a moraine dammed depression on the southeast side of the Mummy Range. It is located in the northcentral area of the

Park. It is fed by the Roaring River out of Crystal Lake, and higher cirque drainages off the Mummy Range. Mummy Mountain is the closest mountain feature to this natural lake. It is located on the east side of the Continental Divide draining into Fall River and then the Big Thompson River.

It was reported to have been named when Abner Sprague brought some British hunters to the lake in 1871, and the color of the lake resembled a green lawn. William Hallett renamed it Forest Lake in the 1880's. It was a popular fishing lake and creek during those times.[3]

An earthen dam was constructed in 1903 by the Farmers Irrigation Ditch and Reservoir Company (FIDRC). The natural lake was originally 16.4 acres in size. The plan was to build a 24-foot-high dam to the spillway height, creating a 47.1-acre reservoir with 811 acre-feet of storage. To release water from the lake, a 3-foot diameter riveted steel pipe was placed with a gate valve on top, and a well shaft to access the valve. Based on surveys the dam height was 20.35 feet to the spillway, with a capacity of 611.96 ac-ft in 1907.[4]

*Lawn Lake July 1976—Dam is to the far left.*

Around 1910, Willard H. Ashton, of the Horseshoe Inn, built a shelter cabin at the lake complete with cooks/caretakers to provide amenities for those wishing to climb in the Mummy Range. He also built a second shelter around 1920, and the Park Service provided a shelter cabin at Lawn Lake until it was razed in 1964.[3]

In September of 1931, the dam was raised four to five feet to 26 feet total height, with 24 feet at the spillway, with a capacity of 817.2 ac-ft. This was never approved by the State Engineer's Office (SEO)[4], nor the Park Service. This addition, which added capacity beyond the acreage allowed by the Park Service, became a trespass issue during the assigning the blame and lawsuit phase after the flood.

The Farmers Irrigation Ditch and Reservoir Company owned and operated the Lawn Lake dam at the time it breached on July 15, 1982.

## Cascade Lake (Reservoir)

Cascade Lake was built on Fall River, just about one mile inside what would become the Park boundary, by F.O. Stanley to provide water for a powerplant he built in 1907-1909 to power his hotel with electricity. He purchased water rights from Pieter Hondius and F.L. Clerc for this use. The dam was enlarged to seventeen feet high in 1923 and was constructed of concrete. It had been dredged twice since 1942 to keep its capacity, the most recent dredging was in November 1981. The lake and dam were acquired by the Town of Estes Park in 1945. The water level was twelve feet deep, the dam was a concrete gravity dam, 143 feet long, and held 12.1 acre-feet of water at maximum capacity.[4] A twenty-inch steel intake pipe provided water from Cascade Lake to the hydro-electric plant. A thirty-inch penstock was added in 1923 to increase the ability to provide more power.[5] The Town had also purchased the hydropower plant from Public Service Company, which operated from 1909 until July 15, 1982.

## Lost Lake (Glacier Reservoir #1)

Lost Lake is about a ten-mile hike from the Dunraven Trailhead on the North Boundary Trail near Glen Haven, in the northeast corner of the

Park. Only the last half of that hike is inside the National Park Boundary. The drainage begins at the north end of the Mummy Range.

In the 1890s, George Simonds had a homestead claim on the North Fork of the Big Thompson River and built a number of sawmills with a camp to support them in the National Forest. By 1909, the sawmills were abandoned, but Fred Sprague, Abner's brother, had a lodge called Sprague's Resort in the area. Lost Lake was a popular destination by tourists making horseback trips to the lake for fishing.[3]

In 1911, partners Lee and Parker, from Johnstown Colorado, raised the level of the lake with a small dam and scooped out the bottom, so they could claim an equal share of North Fork water.[3]

In 1916, the newly founded Rocky Mountain National Park initiated minor trail work, removing timber and conducting trail repairs. In 1932, the Park Service built the North Fork Patrol Cabin, and in 1934, a twenty-man crew worked on the steep section of the trail near Lost Falls, creating "impressive rock walls", as they are described in the Historical Register. These trail improvements were the basis for the Historical Place designation.[6]

In 1931, the dam was approved by the district court to increase the height to sixteen feet above the outlet, with a dam length of 100 feet. Lake Husted (GD#2) water rights were formally abandoned in 1925.[7]

Lost Lake, as Glacier Dam No. 1, was listed by the Park Service as having a rubble masonry dam, with a height of eleven feet, and having a 138 ac-ft maximum capacity.[4]

## Sprague Lake

Abner Sprague sold his Moraine Park (formerly Willow Park) homestead ranch to J.D. Stead in 1904, and later moved to Glacier Basin and started building a new hotel, Sprague's Lodge, at that site. He also built a lake for his guests for fishing. The source for this lake was Boulder Brook Creek. This reservoir had a 13-foot high by 1700-foot-long dam, which Sprague had added to in 1914 prior to the National Park being created in 1915, because Sprague was concerned the Park Service might not let

*Sprague Lake*

him add on once it became a Park. Sprague also built Cabin Lake at this location. Cabin Lake was filled by water from the spillway on the north side of Sprague Lake.

Sprague Lake is thirteen acres in size. Sprague Lodge operated until 1958 and was acquired by the National Park Service and razed within a few years.[8] Cabin Lake's two dams were breached and partially removed by the Park Service after 1958 as well.[5] The former lodge location is now the parking lot for the picnic area. Sprague Lake has three- quarters of a mile of trail around it and handicapped fishing access.

## Lily Lake

Lily Lake (on Colorado Highway 7) was a natural lake of 14 acres when the area was homesteaded in the early 1900's. In 1911 Gordon and Ethel

Mace homesteaded with brothers Charles and Stuart Mace and built several tourist cabins. Their homestead patent was approved in 1917. The 1961 Longs Peak topographical map shows five cabins around Lily Lake, and ten buildings around the Baldpate Inn on the road to the east of Colorado Highway 7. These included stables for Arabian horses that the Mace family raised.

Isabella Bird wrote about it in the fall of 1873. *"From the dry, buff grass of Estes Park we turned up a trail on the side of a pine-hung gorge, up a steep pine-clothed hill, down to a small valley, rich in fine sun-cured hay about eighteen inches high, and enclosed by high mountains whose deepest hollow contains a lily covered lake, fitly named*

Photograph of Lily Lake by William Henry Jackson, 1873.

*"the Lake of the Lilies". Ah, how magical its beauty was, as it slept in silence, while there the dark pines were mirrored motionless in its pale gold, and here the great white lily cups and dark green leaves rested on amethyst-colored water!".*[9] The water lilies described are Yellow Pondlily *(Nuphar luteum ssp. polysepala).*[10]

Lily Lake is located at the headwaters of Fish Creek, that drains into Lake Estes. Fish Creek is about five miles long. Enos Mills spent a great deal of time here since his cabin was nearby in Tahosa Valley, across from Longs Peak. He wrote about a two-year drought at Lily Lake in August 1911, that dried up most of what he described as a ten-acre spring- fed lake, four feet deep at its deepest point. In his account, the lake went almost dry except for some extensive canals dug by beaver. He further describes the beaver lodge occupied by one remaining old beaver and his attempts to help the beaver through the winter by providing him fresh cut green Aspen limbs. The lake refilled in the spring with the melting snow.[11]

The elevation of Lily Lake is 1500 feet above Lake Estes. It had a storage capacity of up to 39 acre-feet of water. In 1913, the State Engineer's office approved plans for a dam for up to 75 acre-feet of water which would increase the lake size to 17 acres. The dam was constructed in 1915 and raised the water level by four feet, with the high point of the dam at ten feet.[12]

In the 1920's, Lily Lake was part of a destination resort called Uncle Joe's Fish Camp.[5]

The cabins around the lake were part of this camp.

Lily Lake was private property until 1989, when it was acquired and transferred to the National Park Service to prevent development. The dam then became part of the NPS inventory of deteriorating dams to either rebuild or remove.

# CHAPTER FOUR

# The Big Thompson Flood
# of 1976

The main Big Thompson River (BT) flows out of Lake Estes below Olympus Dam and heads down the Big Thompson Canyon to Loveland. The highway distance of the canyon proper is about 20 miles. About halfway down the canyon, the North Fork of the Big Thompson comes in from Devil's Gulch Road and the Glenhaven area and joins the "Big T" River at Drake. The normal flow of water in the Big Thompson River is 200 cubic feet per second (cfs).[1]

A number of times prior to 1976, the Big Thompson had flooded since State Highway 34 was completed through the canyon. Storms and flash floods occurred in September 1938, and in June 1941 and 1945. In 1951, heavy rains in the Cedar Cove area, and the breach of Buckhorn Reservoir killed seven people and caused $600,000 damage in Loveland.[1]

One thing residents got used to in the summer in Estes Park and surrounding communities was late afternoon thunderstorms and rain. The "monsoon season" often varied in length between May and July, usually brought by Gulf of Mexico moisture traveling up to the Front Range and mountains of Colorado. Cool air traveling down from Canada collides with this air, often triggering thunderstorms. Moisture builds up over the Continental Divide in Rocky Mountain National Park, and as it works its way east, builds thunderstorms. In the backcountry, the usual goal for mountaineering or climbing above tree line is to be up and back down before late morning so the threat of lightning is less dangerous. Much of my summer each year working on Trail Crew was being soaked at least a portion of each day, and we were often wet for entire ten-day periods while camped out in the backcountry.

On July 31, 1976, I drove down the Big Thompson Canyon to Jax Surplus store in Fort Collins and drove back up around three-thirty or four o'clock in the afternoon. Used to driving a VW Bug, instead I was driving a Ford Galaxy 500 that I thought drove very solidly. Driving a big heavy vehicle like that gave you a feeling of invincibility on the road, a thought I'm sure was shared by many others in the canyon later that night. I was lucky, I was back in Estes Park doing my weekly laundry at the Park Service Utility Area that evening, as it started pouring steady rain.

NPS Rescue Team rangers Larry Van Slyke and Charlie Logan shared with me later, that they watched the massive thunderhead build over the Dry Gulch and Devils Gulch area on the eastern edge of the Park that afternoon. They told me it just sat there for three or four hours while it rained steadily. This was later described by NOAA as a stalled thunderstorm, with moist unstable air meeting a cold front from the north, held in place by weaker prevailing winds, that normally push these building thunderstorms to the east out over the plains.[1,2,3] Estimates of cloud heights as high as 60,000 feet were described with the storm stalled over Hague's Peak in the Mummy Range.[1]

By most accounts, the rain began to fall in the Estes Park area between five-thirty and six o'clock pm, and it was a steady rain for over three hours.[1,2,3] It dumped between ten and fourteen inches on northeastern areas of Estes Park, Glen Haven, Drake and the canyon. Eight inches fell in just one hour.[2] Some parts of Estes Park itself only received about four inches.

With the ground saturated, the water had nowhere to go, but down the canyons. Where the canyon narrowed, the wall of water sped up and went up the canyon walls, described as 20-30 feet high and moving fourteen to fifteen miles an hour.[1,2]

Warnings of flash flooding were broadcast on a Fort Collins radio station as early as 5:30 pm. A Severe Thunderstorm Warning was issued at 7:35 pm[3]. The Larimer County Sheriff had reports of rising water and rockslides at 8:15pm. At 8 pm, The Colorado State Patrol (CSP) received reports of rockslides and a washout on Highway 34. Trooper William Miller responded from near Drake. At 8:35 pm, seven and one-half miles into the canyon, a thirty-foot wall of water washed across the highway and Miller had to swim to safety after radioing in the emergency. He then assisted in warning others about the flood. He survived.

Off-duty police officer Michael Conley was in the canyon with his wife and noticed a rise in the water and a boulder on the road. His wife climbed to safety as he called the CSP to advise them to stop vehicles from entering the canyon. He was lost to the flood after driving further into the canyon, warning people to climb to safety.

Colorado State Patrol officer Hugh Purdy called in the high river water about 8:45pm and was warning residents to get out. At 9:15pm, his last radio transmission was one-half mile east of Drake, as he tried to get out of the flood. His body was recovered eight miles downstream.[2]

The Bureau of Reclamation shut off the release of water from Olympus Dam into the Big Thompson at nine pm.[3] At nine pm, the National Weather Service (NWS) issued a warning of possible flooding in the Big Thompson Canyon, and the first flash flood warning by NWS came at 11pm, long after the flood had left the canyon and was in the Loveland area. The NWS later claimed they had an equipment failure that resulted in the delayed warning.[1]

The floodwall went down the canyons like a massive dam breach, tearing out roads, bridges, propane tanks, and houses, and uprooting trees and boulders. Vehicles, occupied or not, were swept away as well. Floating propane tanks exploded when they hit bridges or rocks. People that thought their vehicles were a safe haven, were not safe. Those that scrambled up the sides of the mountains or clung to trees had the best chance of survival. Many reports of eerie lights seen in the flood waters turned out to be the headlights of submerged vehicles as they were washed downstream, occupied or not.

In Dry Gulch in Estes Park, the peak flow was estimated at 4,500 cubic feet per second, and damaged the "toe" of Olympus Dam, as it joined the main Big Thompson River.[1]

Fox Creek and West Creek feeding into the North Fork, flooded the homes that were closest to its banks, including Glen Haven, before heading on down the canyon.

In Drake, the North Fork of the BT River peaked at 8,700 cfs, and the peak of the main Big Thompson River was 28,200 cfs.[1]

Each of the two forks of the river peaked at different times, given at one and one-half-hours apart. After 20 miles, leaving the canyon past The Narrows, the flow was estimated as high as 31,000 cfs.[1] The Colorado-Big Thompson (C-BT) Project Siphon at the mouth of the canyon was ripped from the canyon walls, it was a pipe nine feet in diameter.

I lived in the old Dave Stirling cabin/studio just inside the Fall River Entrance of the Park, across from Cascade Cottages. As I came through the entrance station, one of my roommates, Tom Olivares, told me that the Big Thompson Canyon was flooding.

When I got to the house, I called Park Dispatch, but they said they had all the rescue team help they needed at that time. I learned that my boss, Jack Gartner, three of our trail crew, and one road crew member, were down in the canyon with the rescue team. I was told to call in the morning to see if more help was needed.

On August 1, I was planning to hike up and meet with my other roommate, Don Bolding, who was the backcountry Fern Lake Patrol Ranger. After calling dispatch and being told I was not needed in the canyon, I hiked in and met Don at Spruce Lake. There were 45-55 National Park Service personnel that assisted with the rescue efforts. As we fished for Brook Trout in the rain all day, we felt helpless as we listened to the Park Service radio traffic throughout the day. We heard rumors of Olympus Dam breaking, which was not true, but mostly we heard confusion as to the true details of the disaster and rescue efforts. Helicopters and ground crews were rescuing and evacuating people all day.

There were four military helicopters in the rescue effort, including two Chinook helicopters, as well as Medivac helicopters, and other civilian helicopters assisting.

Besides the Larimer County Sheriff Department, the National Park Service, and other local first response agencies, Governor Dick Lamm activated the National Guard at 3:30am Sunday morning, with 150 personnel dispatched.

There were an estimated 2,500-4,000 people in the Big Thompson Canyon on July 31.[2] An estimated 2,000 people were reported stranded in the canyon[3], on Monday, Sheriff Robert Watson said that 1,000 people had been evacuated.

On August 2, President Gerald Ford declared Larimer County a disaster area. The American Red Cross and Salvation Army came to aid those being rescued, feed volunteers, and to help with the logistics in evacuation sites.[3]

In summary, the Big Thompson Flood of 1976 destroyed 418 houses, and 152 businesses, with $35-40 million dollars in

*Climb to Safety! In Case of a Flash Flood Sign, near Drake CO.*

damages.[2] Loveland Light and Power lost its power generation plant at Viestenz-Smith Mountain Park. Governor Lamm estimated the road repair for Highway 34 would be 10 million dollars. It was reported that the flood moved boulders that weighed up to 200 tons. There were 144 deaths, including the two law enforcement officers. Many people ran to their cars to escape the flood. 430 vehicles were found in the flood waters. Only 41 (28.5%) of the 144 deaths were canyon residents, 48 (33%) were vacationers from out of state.[1] The remainder were Colorado residents. Approximately 150 people were also injured in the flood.

The loss of the C-BT Siphon pipe caused water levels to drop in Horsetooth Reservoir, so much that remnants of the Town of Stout, buried under the reservoir, were partially visible.[2]

The next several weeks were especially tough on displaced residents. Much of the effort turned to the recovery and identifying those that died in the flood. This was truly a test of how Coloradans responded to a disaster on the eve of their Centennial celebration.

The lessons learned were hard. Communications within the canyon were difficult, phone lines and power were lost. Although amateur Citizen Band radio communications assisted law enforcement efforts throughout the night, most communications were cut off to the majority of people in the canyon. There were no signs in the canyon warning people to climb to safety in case of flooding. Many people tried to escape the flood in their vehicles. A Federal Emergency Management Agency study showed that two of three deaths in flash floods occur when people try to drive out of a flood in their vehicles.[2] The too late National Weather Service warnings were due to equipment malfunction, such as a broken fax machine, and poor communication.

The advent of modern Doppler RADAR weather systems (1980), and early warning monitors, and cell phone technology have improved on the ability to communicate impending disasters. This was demonstrated especially during the 2013 Front Range floods of Colorado, which included another devastating round of floods in Glenhaven, Drake, and the Big Thompson Canyon. "Climb to Safety" signs in the canyon, remind travelers what to do in a flood.

*Authors note:* The 45th flood anniversary gathering in the canyon in 2021 was cancelled due to a flash flood alert issued because of runoff concerns from the Cameron Peak burn area.

# CHAPTER FIVE

# Lawn Lake Flood, 1982

The following notes are from pages from the 1982 Dailyaide calendar notebook
of Henry Schmidt[1], Ditch Rider for the Farmers Irrigation Ditch and
Reservoir Company (parentheses clarify his notes):

**Friday, June 25 1982:**
> *Water Commissioner called*
> *Can close Lawn Lake Gate*
> *Also called Forest (Park) Service in Estes Park on*
> *    accessibility to lake*
> *Report Trail is muddy/snow within 2 miles of lake*

**Tuesday, June 29 1982:**
> *Tel(ephone) Forest (Park) Service*
> *Tel(ephone) Otis White Side*
> *On Condition of Trail to Lawn Lake*

**Wednesday, June 30 1982:**
> *Ranger Wagener (Wagner) Tel(ephone) Trail*
> *    Unact sibel (Inaccessible) for week to 10 days*

**Wednesday July 7 1982:**
> *Telephone*
> *Johnstown*
> *Estes Park*
> *Johnstown*
> *Estes Park*
> *Forest (Park) Service*

> **Thursday, July 8 1982:**
>> *Check (Lawn) Lake, Calvin, Phil and myself*
>>> *Henry Schmidt)*
>> *Calvin pull trailer, Rent one horse*
>> *On order to close Lake Gate.*
>> *Estamat (Estimate) 450 AF (Acre-feet)*
>> *Gauge reading .75 = 12'*
>> *After close of gate*
>> *Gauge reading .03*
>> *Found Broken Key in Lock*

On July 8, 1982, the Farmers Irrigating Ditch and Reservoir company closed the gate valve on the Lawn Lake Dam, reducing the flow out the outlet pipe to near zero. See the previous note from Henry Schmidt. This allowed the lake to fill rather than continue to flow downstream through the outlet pipe.

# The Lawn Lake Flood

### Thursday, July 15, 1982-Rocky Mountain National Park

On a calm, clear morning at eleven·thousand feet elevation, the full Lawn Lake Reservoir dam broke, releasing 674-acre feet of water at 18,000 cubic feet per second.[2] Downstream on the Roaring River, backcountry campsite # 1, a camper was up early, about 5:30am Mountain Time, while his friend still slept.

> *"Went to bed last night about 10:00. We both woke up cause we couldn't sleep. We talked for a while and went back to sleep. This was around 3:00. I woke up early this morning, I woke Steve up, he was too tired to get up. I built a fire and made coffee.*
>
> *I started to hear a sound like an airplane. Also there were loud booms. It got louder and louder. I thought it was breaking the sound barrier. I kept looking for a plane, but couldn't see one. I got suspicious and started to look upstream. I saw trees*

*crashing over and a wall of water coming down. I started to run as fast as I could for high ground. There was a deafening roar. I fell and got up and kept running. I stood on high ground and watched it wipe out our campsite. It knocked everything in its path over. Steve didn't stand a chance. I watched for 15 minutes then started down the Mt. for help."*

         —Steven Cashman, July 15, 1982.[3]

Twenty-one-year-old Steve See was swept away in their orange tent.


*"We woke up at 6:15 and about 6:30 we hear what sounded like a rock slide. We looked out the tent and saw a wall of water heading towards us. We immediately ran for higher ground leaving our equipment. The water did not hit our campsite so we were able to save our equipment which seemed inconsequential compared to the fact that our lives had been spared. We tried to find the other campsite and individuals there but were unable to find signs of it or anyone there. We stayed put until 11:00 am when we came to lower ground and ran into some Rangers and were rescued."-Margaret Brault-Statement taken on July 15, 1982.[4]*

  —Margaret Brault, Albion, MI, camped at Roaring River # 2 backcountry campsite:


*"At about 6:15am I awoke to the sound of trees smashing and water rushing. We looked to the river and saw trees and water rushing by. We went to the lower campsites and saw that everyone was fine. We went back to our campsite and saw Wheelers trail crew. They yelled across the river to stay still and wait. Some time later they came down the other side of the river. They collected people from the lower campsites and we went to a spot where the helicopter could pick us up."*

  —Steve Wilde, Glenview, IL, camped at Cut Bank backcountry

                  campsite:[5]

*"I woke up at 6:15 and the ground was shaking. I heard some trees braking. I got out of the tent and saw a wall of water rushing down the river. I watched the water for xxxx till the ranger came. I saw them across the river. They told us to stay there. They came about two hours later. We went to a clearing and waited for the helicopter. Then they took us to the Ranger Station."*

—Tim Wood, camper:[6]

In Horseshoe Meadows at the Lawn Lake Trailhead, Steven W. Gillette, of A-1 Trash Service, was picking up trash early.

*"At approximately 6:18am, arrive at the Lawn Lake trailhead to empty 4 cans. Getting out of my truck I hear a very deafening noise. Looking I see dirt in the air. Decided that a jet was crashing and to go up the road to confirm.*

*About 30 yards from Roaring River Bridge I notice some water and debris (logs and limbs) on road. Stopping and looking uphill I see trees and rocks being thrown into the air.*

*Truck in reverse to the Endo Valley inter closure gate-turn around to go to trailhead emergency phone. Notice Blazer (Michigan plates) entering off 34. Decide to block road and run to phone.*

*6:22: Call dispatch—Ask what's going on up Endo Valley? I think a lake or something just burst. There is a massive landslide—something coming down the mountain- I have the road blocked. I'm the A-1 Trash Driver. Get (Ranger) Dan Davis down here. Not going to let anyone up the road til Dan gets here.*

*Ranger Shultz arrive where I have my truck blocking access. Shultz and I proceed to the Endovalley winter closure gate. He asks me to lock it. He is on the radio. We both then go to the bridge on 34. While there, I see much more water and debris. After getting to the bridge I see a knee high wall of water*

*and debris coming to the bridge. I tell Shultz that I will block*
*the road with some blockades that I have observed up the road.*
*I do this and by the time I finish-water is coming over the road.*
*- I go on with my business."*

—Stephen W. Gillette, handwritten statement on July
22, 1982.[7]

Where the Roaring River came down Bighorn Mountain, it tore
a deep gash in the hillside, before filling Horseshoe Park with water.
Formerly cryptic Horseshoe Falls became very visible as it crashed
down the mountainside, creating a wide alluvial fan of rocks and trees
and debris at the bottom of the mountain in Endovalley. The alluvial fan
scar is very visible across the valley from locations on Trail Ridge Road.

*Lawn Lake Alluvial Fan, July 10, 2013*

Cascade Cottages were a private property inholding about a mile inside the Park from the Fall River Entrance Station. It had 14 cabins that were rented out in the summer, and they were adjacent to the Cascade Lake and Dam. The Park Service notified the owners, and the occupants were evacuated to higher ground before the flood waters reached Cascade Lake.

Mr. L.V. Davis, owner of Cascade Cottages and Mrs. Iris Turner, a volunteer for Mr. Davis were interviewed about the Cascade Dam failure:

*Mr. Davis stated that at approximately 0630, Thursday morning, he was warned by a Park Ranger that the Lawn Lake Dam had broken and flood waters were headed their way via the Fall River. After warning his guests of the situation and proceeding to safe ground, he stated that he watched the water level begin a visible rise at the Cascade Dam at about 0715. He commented that the water level eventually reached 6 ½ to 7 feet above the top of the dam. About 45 minutes after the first wave of water hit the Cascade Dam, there was a sudden drop in the water level going over the dam. Although the dam itself was no longer visible, it was at that time, approximately 0800, that he believes the Cascade Dam gave way.*

*Mrs. Turner stated that at about 0700 she saw two Park Rangers near the dam. They both left the cottage area very quickly. Shortly thereafter, she saw about ½ foot of water spilling over the top of the dam. She stated that as the water level increased, the utility house on top of the dam held until the point at which the dam appeared to rupture, and it was washed downstream. At that time, she looked at her watch which read 0759.*

—Dr. Davis and Mrs. Turner statements taken on Tuesday, July 20, 1982, by Park Technician Lisa C. Nichols.[8]

*Cascade Dam Failure*
*right: Water overtopping the*
*Cascade Dam. Photos courtesy of*
*G.D George*

Ingrid Thomas of Arlington Texas, describes what she saw in Aspenglen
Campground, approximately one-half mile downstream from Cascade
Dam along the Fall River:

> *"My daughter Bridget and I were camped at Aspenglen Camp-*
> *ground at the walk-in site # 76. On 15 July, 1982, we were*
> *walking back to our campsite from the toilets-about 7:00*
> *AM-when we met a girl with bedding under her arm on the*
> *first bridge, and she asked us how much time we had. My reply:*
> *"Time for what?" She answered "The ranger told us to evacuate*
> *our campsites because a flood was coming." We did not find a*
> *note from the ranger at our site, but we started packing sleeping*
> *bags and personal belongings. Bridget walk(ed) a bit ahead of*
> *me, and when I crossed the first bridge, Bridget was walking*
> *back to get a second load. This was the last time I saw her.*
>
> *I briefly want to describe the condition of the flood. First,*
> *about 2 min. before I left our site, a noise such as thunder was*
> *heard. We saw people running away from the creek below us.*
> *As I got a jacket from the picnic table, water slowly, then swiftly*
> *flowed under the table and within seconds rocked our cooler. 20*
> *yds from the site water was splashing about my sneakers, and*
> *past the first bridge I crossed, I had to watch where I stepped.*

*Thunderous water was at the first bridge. All of us were stunned to watch this spectacle in awe, fear and without any knowledge of what to do. I did not find out for many minutes that a dam had broken. I attribute Bridget's death to lack of information to us and misjudgment of the magnitude of the situation on her part."*

—Ingrid Thomas statement dated 17 July 1982.[9]

Rosemary Coates of Peoria Illinois describes what happened in Aspenglen Campground:

*"We arrived at Aspenglen Campground Sunday around 10 o'clock (Site 68 I believe) -in front of the bathrooms. We toured the Park-Trail Ridge-Bear Lake. We were getting ready to leave for Colorado Springs. Terry and I had already taken down a large brown awning type tarp we had over our table. A girl came out of the walk-ins and said "Do you have to leave too?" I said yes and then after I thought about it I said "What do you mean have to, we are just getting ready to leave for Colorado Springs." She said a ranger had been by to tell them they had an hour to leave those sites as there was a danger of flooding (no mention was made of a dam breaking).*

*In about 10 minutes we heard a rumbling noise and saw the water in the river rising. Terry said to wake the kids up so they could see, and he (page one of her statement ends here......Terry Coates was last seen crossing the first bridge by the walk-in sites about 07:55am with an unidentified man" (reported missing as well).*

—Rosemary Coates statement on July 17, 1982.[10]

Aspenglen Campground, just inside the Fall River Entrance to the Park, was a campground with car and camper vehicle sites and ome walk-in tent sites on river islands in the campground. There were

275 campers in Aspenglen on the morning of July 15th. Park Rangers had first issued evacuation orders at 6:50 am, and final evacuation orders at 7:23 am. Waters overtopping Cascade Dam may have reached the campground as early as 7:33 am, but at approximately 07:42 am, when the dam failed, a surge of water hit the Aspenglen campground with deadly force within five minutes. That surge was estimated at 16,000 cubic feet per second.[2]

After the water left Aspenglen Campground, it continued down Fall River towards Estes Park, wiping out the State Fish Hatchery and killing 90,000 fish. It also completely ruined the Stanley Powerplant. The flood washed away or

*top: Lawn Lake Flood debris and mud in Estes Park. Photo by U.S. Bureau of Reclamation; above: Estes Park looking north, Town Park/Library, center. Photo by* Loveland Reporter-Herald.

damaged cottages and cabins, Fall River bridges, and entered Estes Park about 8:15 am. As the water, mud, trees and debris entered Estes Park, it was less forceful, estimated to be six feet deep, but filled most of downtown Estes Park businesses with this sediment.

The flood waters entered Lake Estes, contained by Olympus Dam, and by 9:30am, the flood had completed its destruction. At 7:12 am, the Bureau of Reclamation (USDI BOR) had been contacted about the ability of Lake Estes to contain the flood. Since Lake Estes was four feet below its maximum capacity, the BOR determined the flood could be contained. As a precaution, they shut off the water coming from the Western Slope through the Alva B. Adams tunnel. Lake Estes rose two feet as a result of the flood.[2]

From the *Calendar Notebook* of Henry Schmidt, Ditch Rider for the Farmers Irrigation Ditch and Reservoir Company on Thursday, July 15, 1982, was noted at the bottom of the page under payroll time notations:[1]

**Lawn Lake Dam Broke.**

# Aftermath of Lawn Lake Flood,
## July 15, 1982

Three people died in the flood, Steven See, at Roaring River campsite #1. Bridget Dorris, and Terry Coates, were lost in the waters at Aspenglen Campground. A fourth person was observed being washed away with Terry Coates in several witness accounts from Aspenglen Campground, but this was never corroborated, nor evidence found documenting this "missing person" any further. This missing person was described below.

From Supplementary Case/Incident Record (CIR) 820603, Page 2 of 4: written by Ranger Charles E. Logan, 7/19/82:[11]

*4) ADULT MALE, identity unknown*
*Description: approx. 35 yoa, possibly wearing light colored clothing (blue? grey? tan?), baseball cap (blue/white with writing?)*

*Last seen: with TERRY COATES crossing first bridge into the walk-in sites at Aspenglen Campground (RE: witness report, form 10-344, 07/16/82, Rosemary Coates). Estimated time of 0755 base on reasons stated above.*

The flood took out two dams, at Lawn Lake, and Cascade Lake. It destroyed the Stanley Powerplant, and the State Fish Hatchery. The flood damaged 177 businesses in Estes Park, about three-quarters of the town. One-hundred-eight residences, and thirteen bridges were also damaged or destroyed. An estimate of the damages to the town exceeded $31 million dollars.[2]

The National Park Service located and flew out backcountry campers and interviewed witnesses after the flood, and most were hand-written accounts or typed by the interviewer onto Supplemental Case Incident Records, including those included earlier in this chapter. There were 24 permitted backcountry campers, and four illegal campers. As events unfolded, and searches continued, so did details of the flood.

*"I camped the night of 7/14 at Lawn Lake site # 5. A loud
roar woke me up in the early morning. Got up and saw dam
destroyed. Saw 4 persons 1 man age 50, 1 adult woman, 2
teenage women hiking in 7/14, destination "Bighorn" 7 ½ mi
in. Talked to the man. He said didn't think could make their
designated camps, so would camp "wherever"—last saw them
in switchback area about 11:30 am."*

—David Allen Box statement, Supplementary Case Incident
Record July 15, 1982. Laurie Shannon, Interviewer.[12]

*"I arrived at Lawn Lake Patrol Cabin Monday July 12, 1982 at
9:30 pm, and stayed until approximately 3:00 Wednesday, July
14. During that time I was making observations of the Bighorn
sheep as part of the research for Rocky Mountain National Park.*

*Since I was not interested in the lake or dam I did not
spend much time observing the lake, dam or spillway.*

*Wednesday morning, around 8:00, where I filled my
water bottle from the spillway, I noticed that it was full and the
volume of water was that of a river the size of the spillway. It
was approximately a foot deep. The lake was full and, with the
wind from a front that passed through, the waves were breaking
on top of the dam. This lasted about an hour. As to the level of
the water on the dam and any seepage or flow of water below
the dam, I did not notice anything at all. Other than a rain
Tuesday afternoon there was no significant amount of precipitation during the time I was there. The ground was soggy
around the lake from the runoff and the frequent rain storms
this summer."*

—Donay Hanson, Biological Aide, NPS, statement:[13]

Bert McLaren, Park Ranger, NPS, also wrote a post-flood statement that the waters in Lawn Lake were full on July 13, 1982, prior to
the morning of the flood. He was hiking to Mummy Pass when he was
at Lawn Lake and made his observation.

Another camper at Lawn Lake described waking up at 2 am on July 15, hearing a noise that she thought was wind rustling through the trees. When she looked out, she saw that the trees were not being blown by the wind. Likely she heard or felt the beginnings of the dam breach. The Lawn Lake campsites were above the dam site, and to the east of Lawn Lake on higher ground. Other campers also described loud noises such as a roar before the breach occurred, between 2am and 4am.[15]

The National Park Service's initial response during the flood was warning downstream authorities, blocking roads, and monitoring the progression of the floodwaters from Horseshoe Park, down Fall River, Cascade Dam, and warning residents and guests at Cascade Cottages and campers in Aspenglen Campground. The access road bridge to Aspenglen was cut off and full evacuation of the campers didn't occur until a temporary road was built around midnight on July 15th. Most of the campers remained at Aspenglen until the following morning.[2]

Immediately after the flood, the National Park Service began searching or accounting for all persons known to be in the flood zone. Fortunately, the backcountry permit system had registered all legal campers in the potential backcountry flood zone, those down river from Lawn Lake. Aspenglen Campground registrations allowed for an accounting of those 275 individuals as well. Trail crews and Rangers began searching for those listed as being in the flood zone. Ranger Charles Logan reported 43 Park Service personnel involved in search and rescue efforts.[11]

The Larimer County Sheriff and Town of Estes Park Police covered the Fall River areas downstream and outside of the National Park, between the Aspenglen Campground and the Town of Estes Park.

The warning phone call from Steve Gillette at the Lawn Lake Trailhead allowed for adequate notice to help authorities contact and evacuate those in the potential flood path. KSIR Radio broadcast information pertaining to the timeline of the flood beginning at 7:05 am, which also aided Estes Park residents and business owners in preparation for the coming floodwaters.

Later, on the day of the dam failures, The Western Area Power Administration (WAPA) provided a helicopter and pilot to the Bureau of Reclamation in Loveland Colorado. The helicopter and crew flew to the Lawn Lake Dam site around 9:00 am. They filmed color movie footage of the dam site, the flood path down the Roaring River drainage, and the flood through Estes Park. The crew also took 48 color photographs documenting the dam breach and the flood path.[2]

From a Memorandum written by James D. Harpster, Public Affairs Officer, Rocky Mountain Regional Office:[14]

> *"Upon learning at the Regional Office of the Lawn Lake dam failure on Thursday, July 15, 1982, five representatives of the RO (Regional Office) drove to Estes Park and Rocky Mountain NP (National Park) headquarters that day."*

Mr. Harpster and four others from the Solicitor's Office met with Superintendent Chester Brooks and Assistant Superintendent James Godbolt and Chief Ranger Dave Essex. In the memorandum, Mr. Harpster talked about logistics including the chartering of two helicopters for rescue and recovery work, and the Army Corps of Engineers working on bringing in temporary bridges for providing access to areas cut off by the flood and road failures.

A discussion occurred about the last inspection of the Lawn Lake Dam in August 1978 by the State Engineers' Office.

Harpster further stated that in a phone call to Mike Baugher about 1p.m.,

> *"Baugher advised that our Washington office reported an inspection by unidentified Soil Conservation Service employees on April 30, 1981." Further, Baugher said an individual had called him during the morning to say he (the individual) was at the dam on Thursday, July 8, and had observed three "Farmer Types" struggling with a valve, cursing and fuming about its stubbornness. They told him, he said, that the valve was "tricky".*

This statement corroborates the daily notebook entry of Henry Schmidt, of the Farmers Irrigation Ditch and Reservoir Company of the closing of the gate valve at Lawn Lake on July 8, 1982, cited earlier in this chapter. Three ditch company employees were present at Lawn Lake on that date.[1]

Following the meeting with the Park Service Leadership, Public Affairs Officer Harpster and members of the Solicitor's Office went to the helipad, talked to backcountry campers that had been evacuated, and then flew to the flood zone. After flying up the Roaring River drainage to Lawn Lake, Harpster describes seeing three individuals at the dam, "*apparently tape measuring the dimensions of the break in the dam*". He also describes "*A large metal corrugated pipe was in evidence in what appeared to be the bottom of the break.*"

Immediately after the dam failure on July 15th, State and Federal agencies were notified. The State of Colorado Division of Water Resources, Dam Safety Branch, State Engineers Office (SEO), sent a team of two engineers to the Lawn Lake Dam site, and two engineers to the Cascade

*Lawn Lake outlet pipe in bottom of dam breach. Photo 7-20-82, Rocky Mountain National Park, #130772-1*

*Aerial photo of the Lawn Lake Dam Breach. Photo 7-20-82, Rocky Mountain National Park # 130762-7*

Dam site. The Bureau of Reclamation (BOR), the Western Area Power Authority (WAPA), U.S Army Corps of Engineers (COE), Federal Emergency Management Agency (FEMA), and the Federal Energy Regulatory Commission (FERC), all responded to the disaster. WAPA had already provided a helicopter to ferry inspection teams to the site and fly stranded campers out.

The National Park Service conducted searches to recover the victims within the flood path in the Park and recovered Steven See's body in Horseshoe Park at 12:09 pm on July 16th. After his recovery, the Park Service concentrated their search and recovery efforts to areas of Aspenglen Campground and downstream.

Beginning on July 17th, the Park Service in cooperation with the Larimer County Sheriff's Department searched areas below the Park toward Estes Park for the remaining missing victims.

The bodies of Bridget Dorris and Terry Coates were recovered about one to 1.5 miles downstream of the Aspenglen Campground, on July 20, 1982. The "fourth victim" was never found.

On July 16, the day after the flood, water quality became a problem in the Estes Park area. Raw sewage from the Estes Park city sewer plant was entering Lake Estes, due to mud and sediment in the treatment plant. Downstream water users were warned about water quality, and after four days, water quality improved. The Bureau of Reclamation contracted $36,717 for Lake Estes reservoir debris removal.[2]

The path of the flood from Lawn Lake to Elkhorn Avenue in Estes Park was twelve and one-half aerial miles along Roaring River and Fall River.[15]

A Presidential Disaster Declaration for Larimer County was announced on July 22, 1982.

# CHAPTER SIX

# After the Flood, Liability, and the Blame Question

F̶ollowing the flood, questions of liabilities for dam safety and inspection were brought up by the public and the media. The Colorado Legislature had previously passed a bill in 1973, where dam liability was legislated to be the responsibility of the dam owner.[1,2,3,4]

In 1981, Colorado Senate Bill 259: Reservoirs—placed a limitation of liability for damages; and provides that employees, shareholders, and board members of privately owned reservoirs are not liable for damages caused by leakage or overflow or by floods caused by a breakage of embankments if the owner has in effect a policy insuring against such damage in minimum amounts of $50,000 per claim and $1,000,000 aggregate losses in one incident. It further specifies that the policy need not cover acts or omissions which are fraudulent or criminal and that there is no exemption for liability if such acts, or ultra vires acts, are involved.[3] This sets the stage for the outcome of many of the lawsuits that followed the 1982 Lawn Lake Flood.

Governor Dick Lamm placed the blame on the ditch company and disavowed any blame for the State Engineer's Office and the State of Colorado.[1,2,4]

## Lawn Lake Failure

The Governor asked the head of the Department of Natural Resources (CDNR) which includes Division of Water Resources and the State Engineer's Office (SEO), on July 16, to review the Lawn Lake failure and determine what could be done to prevent future incidents. He asked for a determination within seven days and asked five questions relating to dam safety, inspections and hazard status. The SEO took the lead for the investigation of the Lawn Lake failure; however, many federal and state agencies were involved in this work with two site visits after the day of the failure, on July 17 and July 22. These investigations were extensive, and were outlined in interim reports of the State Engineer's Office until two principal causes of failure were identified.[1]

The dam was 560 feet long at the time of the breach, and 26 feet high, the water level was 24 feet full, spillway height, at capacity with 674 acre-feet of water. There was a twelve- inch layer of organic material four to five feet below the top of the dam, likely added when it was enlarged in September of 1931. There were possible "marmot burrows" at the 3-4-foot level. The only recent rain was on July 10, with 0.11 inches of rain that fell.[1]

The breach dimensions were 97 feet at the top of the dam, and 55 feet at the bottom of the breach, which was 28 feet deep.

Six possible causes of failure were investigated by the SEO:

1. Overtopping
2. Earthquake damage
3. Rodent Damage
4. Frost Penetration

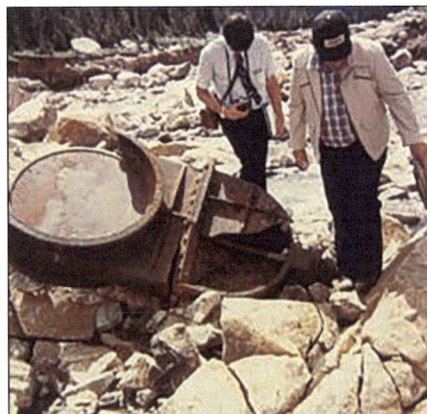

*left: Outlet valve during construction, 1903. Photo courtesy: Investigation of the Failure of Lawn Lake Dam, State Engineer, State of Colorado, 1983.*

*right: Outlet valve-Found below Lawn Lake Dam breach. Photo courtesy USGS.*

5. Embankment Stability

6. Piping (failure)

The Farmers Irrigation Ditch and Reservoir Company (FIDRC), through their insurance company, also hired a local engineering firm that made a site visit on July 20, 1982 to investigate the dam breach.

As a result of eliminating other failure modes, the dam breach investigation focus came down to the outlet valve and the connection between the valve and the outlet pipe.[1,4,5,6,7]

This valve was placed during the original construction of the dam in 1903. The connection of the valve was cold-caulked Lead in 1903. The specifications called for it to be encased in "concrete for three feet above the pipe and two feet below the pipe and one foot thick. However, concrete encasement was not done during construction. The Lead caulking seal was also intended to be placed using hot fluid Lead.

When this caulking connection deteriorated, it led to erosion within the earthen dam from water seepage, that led to the dam breach 79 years later. In a dam site inspection on September 2, 1982, a metal detector was used to gather a large volume of Lead from the site to corroborate this finding.[1,4,5]

This drawing depicts the design and placement of the outlet and gate valve in the Lawn Lake Dam.[6]

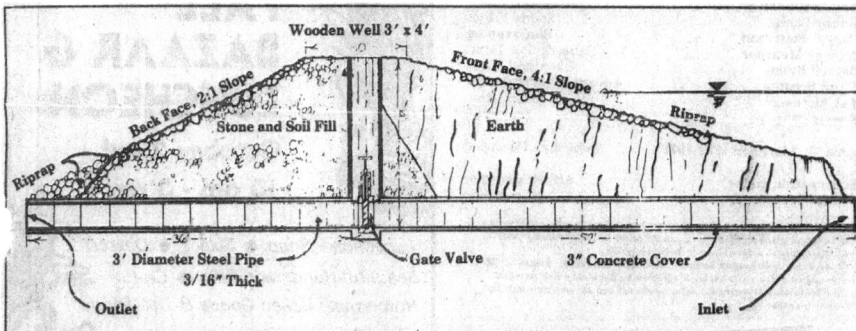

This schematic of Lawn Lake Dam indicates the placement of a gate valve, which is believed to have failed, resulting in the July 15 inundation of Estes Park. The drawing depicts 1902 construction and does not reflect a raise of 10 to 12 feet which was done without authorization by the dam owner, Farmer's Irrigating Ditch and Reservoir Co. of Loveland.

*Drawing of outlet valve-Lawn Lake Dam, courtesy of Estes Park Trail-Gazette*

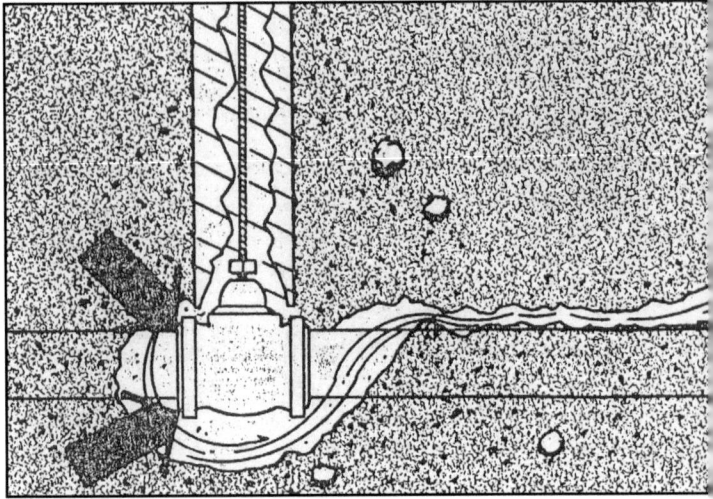

*Diagram of "Piping" at Gate Valve Junction. Photo courtesy of The Denver Post.[7]*

The valve-failure finding by the State Engineer's Office was released on February 14, 1983.[1]

Piping is where erosion occurs from a leak that works away at the dam from inside until the erosion eventually becomes a breach causing failure of the dam.

*"Such a scenario of leakage, piping, and resulting embankment failure required the prerequisite conditions of a closed or nearly closed outlet gate and a near-full reservoir, to supply the head (pressure) necessary for accelerated leakage and subsequent progressive piping through the years. Just prior to failure, the outlet gate was nearly closed and the reservoir was full."*

*"The void created along the outlet pipe would have become enlarged by the internal erosion of embankment material. Campers in the vicinity of Lawn Lake the night before the failure reported a noise that sounded like strong winds. The sound indicated that the reservoir may have been discharging through the dam for at least 3 to 4 hours before the hole enlarged sufficiently –or embankment failure occurred—above the outlet pipe, causing a total breach of the dam where the outlet works were located".[5]*

The report on the dam failure prepared by the engineering firm hired by the dam owner's (FIDRC) insurance company was provided to the FIDRC on January 1, 1983, and to the SEO on February 6, 1984. It reported that the failure "probably" was the result of high seepage pressures from the reservoir, with sloughing ultimately causing the embankment crest to fall below the level of the reservoir.[4]

## Cascade Dam Failure

The Cascade Dam was overtopped for about 40 minutes before it failed. The failure was reported not as a "breach" in the dam, but the gravity dam actually tipping over from the excessive force of the water flow.[5] An SEO internal memo reported that the dam failed due to hydrostatic forces of overtopping combined with erosion of the abutments, causing the dam to topple.[4]

# Lawsuits over the Lawn Lake Flood

Beginning almost immediately in August 1982, lawsuits over the flood began being filed in court. Senate Bill 259 in 1981, limiting dam owner liability was passed as Colorado Revised Statute 37-87-104. Farmers Irrigation Ditch and Reservoir Company had an insurance policy of 1.4 million dollars. The State of Colorado and Town of Estes Park liability, along with the ditch company, only had a total of 2.2 million dollars available to loss holders.[4]

Many insurance companies held that the flood was not an "Act of God", or natural disaster, and because it was a human-caused failure, refused to pay claims by residents and business owners.[4]

The Estes Park Action Committee, headed by Odd Lyngholm, asked "Super-Lawyer" Gerry Spence to examine post-flood legal resources. Spence, a Wyoming lawyer, having never lost a case, was considered the best trial lawyer to help recover flood losses by business and private property owners. The headline in the Estes Park Trail-Gazette read,

*"Country Lawyer Spence charms crowd of 400"*, on July 30, 1982. However, his "fee" of 50% of judgements and settlements, also meant fewer dollars would be recovered for the victims of the flood. This resulted in eventually dropping Spence from these flood cases.[8] *"Country Lawyer Spence fired"* was the final headline in the Trail-Gazette regarding his services.

One of the earliest lawsuits filed on July 26, 1982, was the Fall River Valley Mobile Home Park (Filbey), against FIDRC and the State of Colorado for 2.5 million dollars. Restaurant owner Nick Kane and three others followed with a claim for 3.25 million dollars in damages against FIDRC and the State of Colorado. The list of lawsuits grew and continued in 1982 through 1984, for a total of $35,638,503 in claims against the FIDRC ditch company, State of Colorado, Town of Estes Park, and National Park Service[6]. With only 2.2 million dollars in funds available for relief, the claims ended up only paying about six cents on the dollar.[4]

Many of these lawsuits were directed at the dam owner. The State Engineer's Office also was scrutinized for "negligence" due to the infrequency of dam inspections. The Town of Estes Park was included in lawsuits due to the ownership of the Cascade Dam. Finally, the National Park Service was sued, alleging that they were aware of the Lawn Lake dam's deteriorated status, and allowing the FIDRC to "trespass" on Park Service land with their dam expansion in 1931. They were also sued for negligence for the death of Terry Coates in Aspenglen Campground.

In early 1983, Senate Bill 199 21-9, proposed removing the State of Colorado's immunity from negligence for the dam failures, and proposed providing a fund up to $10 million dollars for flood victim relief. The bill was unsuccessful and was withdrawn from the legislative session.[9,10]

Governor Dick Lamm said that the legislature made a mistake in 1981 passing a law absolving owners of private dams from personal liability if they carried $1 million in liability insurance.

Many of the lawsuits were unsuccessful or drawn out in their conclusion. That frustration was seen in the headlines and newspaper articles over about 12 years after the flood. In the article, *RMNP Wrongdoing alleged in Flood Lawsuit*, January 29, 1984, 29 Estes Park residents and five insurance companies sued, alleging that the Park Service should have known the dam was deficient, since their plan was to buy them out and eliminate them. The Park Service argued that the dam maintenance was out of their hands based on Colorado law, and given to the dam owners and operators as to liability.[11]

In *Developers file $5 Million Claim Linked to Lawn Lake Dam Break*, July 4, 1984, a federal Judge ruled that the claimants must substantiate their claims, with a liability of only 1.5 million dollars in insurance by FIDRC.[12]

Other headlines demonstrate the frustration with the legal battle:

*Flood Victims must pay for Faulty Lawsuit*, March 11, 1987. The judge ordered the 19 victims to pay $15,263, because they wrongly pursued the newly reorganized ditch company, while the FIDRC was already in District Court over the flood.[13]

The Office of the Solicitor, in Denver Federal Court, on July 10, 1987, dismissed the Plaintiffs' causes of action based on the theory of blasting too close to the Lawn Lake Dam and the theory of common law trespass when the water escaped from the dam onto the Plaintiff's property. They also answered to cause by trail maintenance activities in the area, dismissing that as well. The Office also stated that helicopter access to the lake is prohibited by Federal Regulation except for official government business.

The frustration by flood claimants continued as the years passed, headlines such as, *1982 Flood Claims Remain Unresolved*, published on May 13, 1988, continued the debate.[14]

In the article, *Court Spikes hopes for Lawn Lake Flood Claims*, July 19, 1991, the U.S. Court of Appeals upheld the denial of a 25-million-dollar damage suit. The article stated that this ruling could end compensation attempts for 200 business owners under Attorney Jerry Cardwell. The court said there was no proof that the government contributed to the flood or added to the dam's danger because of regulations that hindered Park Service management.[15] The Case was Aldrick Enterprises, Inc. et al. versus the United States of America (NPS).

In the article, *Absence of Proof Relinquishes the Park Service of Dam Liability*, July 19, 1991, Tim Asbury writes that the court cited a 1973 State Law making the reservoir owner liable for all damage from floods caused by ruptured reservoirs. The court said limited access to the dam as required by RMNP to preserve environmental values was not a convincing factor of ownership interest.[11]

A judgement was made in the case of Rosemary Coates against the (United States) National Park Service in February 1985, awarding $480,000 to the family of Terry Coates. The award represented 60% of a potential award amount, holding the victim 40% responsible for his own death, since he had been warned of the flood and crossed back into the river area.[4,16] The federal judge cited reasons for the award:

Park Rangers erred when they thought that the flood would be contained or dissipated in Horseshoe Park, upstream of the campground.

The Park Service (defendant) failed to post an observer at the Cascade Dam to monitor the flood waters and their impact on the dam.

They failed to have a ranger in Aspenglen Campground at all times during the flood.

They failed to give adequate warning of the danger resulting from the failure of the dam to Terry Coates.

They failed to maintain or implement a plan for dealing with emergencies, such as the one involved here.[16]

# CHAPTER SEVEN

# Dam Removal or Repair?

T he Dam Safety Program and frequency of dam inspections in
Colorado was immediately under scrutiny following the Lawn
Lake Flood. With an estimated 2250 dams in the state, and only seven
dam inspectors, the Colorado Program was underbudgeted and under
fire. All government agencies responsible for dams, including the
National Park Service, were under pressure to examine and inspect
remaining dams under their jurisdiction.

In an Internal Alert Memorandum to the National Park Service
Director, dated three weeks prior to the Lawn Lake Dam breach and
subsequent flood, two other dams in Rocky Mountain National Park
were listed as posing serious hazards, Pear Lake Reservoir and Sandbeach
Lake in Wild Basin. Lawn Lake was not mentioned in this internal report.[1]

Even with the flood event just past, FIDRC was still interested in
rebuilding the Lawn Lake dam. Dan Campbell wrote an article on July
23 1982, in the *Trail-Gazette* entitled *Ditch Company Still Unsure about
Dams Future Status.*[2] Ray DeGood, the Attorney for FIDRC emphasized
the right of the company to store 612 acre-feet of water at that site. With
20 stockholders in the company, 30 share stocks, and five members on
the board, they felt they had an obligation for those water rights.

In a series of newspaper articles in the *Estes Park Trail-Gazette*,
reporter Jackie Hutchins writes about the remaining dams in Rocky
Mountain National Park following investigations and inspections by
the National Park Service and others on July 22. [3,4.] The six dams listed
include Pear Reservoir, Bluebird Reservoir, Sandbeach Reservoir, Lost
Lake Dam and the Sprague Lake Dike and Lawn Lake. The Lawn Lake
site and downstream areas were extensively mapped, measured and

investigated by multi-agency teams. The final determination of what caused the Lawn Lake breach would come from Jeris Danielson of the State Engineers Office. One reservoir not included in this list is the Lily Lake Dam, which was only acquired by the Park Service in 1989. A later article, entitled *Park Inventories Remaining Dam*, discusses Snowbank Dam and the Lost Lake Dam.[5] The Snowbank Lake Dam is a very shallow dam that would present no danger, even if it were to break, according to Park Superintendent Chester Brooks. The dam at Lost Lake was examined and found to be in good condition, according to Brooks. Both were determined not to pose a hazard downstream.

Results of the inspections were reported on July 28, 1983, and Sandbeach Dam and Pear Reservoir Dam were found to be leaking, and work was being done to drain them to reduce the hazards. Sandbeach was the only reservoir of the three that was currently being used to store Longmont-owned water. Work on the Sandbeach Lake spillway was being undertaken by the City of Longmont, due to a leak in the rock and mortar construction of 4-5 cubic feet per second, and an evacuation of areas in Wild Basin below the dam was done while the work was being completed. Sandbeach Creek, which drains the lake, comes into the North Saint Vrain Creek just below Copeland Falls. Pear Reservoir, which had water 18 inches below the top of the dam, was being drained by eight inches per day to reduce the hazards. The City of Longmont was also working on removing debris at the dam at Pear Reservoir. After completion of lowering the water levels, the outlets were left open to reduce the hazards. Bluebird Lake was already at its natural lake water level below the dam.

Park Service officials were also beginning work with state officials to try and acquire the development rights for Lawn Lake, according to Park Information Clerk Jeff Karraker. If they get development rights, the dam won't be rebuilt.[6]

However, a moratorium on acquisitions had been placed by then Interior Secretary James Watt, which includes the Wild Basin reservoirs and the Lawn Lake site.

The Wild Basin reservoirs were owned by the City of Longmont, and Longmont hoped that the Park Service would buy the reservoirs. Copeland Lake, also owned by the City of Longmont, was included in a Park boundary change that would have allowed the lake to be taken out of the Park, but that was still under negotiation between the City of Longmont and RMNP.

Since the City of Longmont had an alternate storage site at Button Rock Dam/Ralph Price Reservoir, the waters from these three Wild Basin Reservoirs could be contained in an out of park reservoir downstream.

On August 6, 1982, a multi-agency task force created a Flood Hazard Mitigation Report, released by FEMA. In this report, the NPS agreed to pursue the acquisition of the water impoundment rights to Lawn Lake and three other dams in the Park. These included Sandbeach, Pear, and Bluebird Lakes (Arbuckle Reservoirs). Lost Lake (Glacier Reservoir #1) water rights were already owned by the NPS.

Arguments over the Colorado Dam Safety Office went back and forth for months, and the office was examined by the Colorado Legislature beginning in August 1982. Questions about scheduling of dam inspections, frequency of dam inspections, number of inspectors needed, were all scrutinized. Eight additional inspectors were requested, but only half of those positions were funded. The desire to inspect all 2250 dams annually in Colorado was not a reality, only the high-risk dams, based on risks to persons and property would receive higher priority.[1]

The Coloradoan newspaper reported on a revised Dam Inspection Plan bill in the Colorado Legislature in March 1983.[7] This bill did pass, here is the summary from the 1983 Digest of Bills:[8]

**House Bill 1416: Reservoir safety:** duties of state engineer. This bill increases the number of reservoirs subject to the jurisdiction of the state engineer by reducing the minimum size of reservoirs which require the state engineer's approval for new dams, alteration, enlargement, modification, or repair, other than normal maintenance and repair. It requires the state engineer to give written notice to counties and cities downstream from any dam on which a complaint has been filed, and to report on the steps taken to resolve such complaints. Such notice must also be given in any case in which the state engineer is required under his rules and regulations, to approve specifications for construction or alteration. It requires the state engineer to rule on applications for proposed dams and reservoirs within specified time limits and requires such determination to take into account the hazard classification of the proposed dam, as may be established by rule and regulation. It increases maximum fees collected by the state engineer for inspection and engineering services.

It provides for the exchange of information compiled by the state engineer concerning high hazard dams with counties and cities which could be affected by dam failure, and requires the state engineer to furnish detailed reports on such dams on or before September 1, 1983. It allows cities and counties to be charged $125 for such reports. It requires reservoir owners and operators and cities and counties to assist the state engineer with any additional information he requests based on such report by specified dates, and provides for an updated report. It requires cities and counties receiving a report to conduct a review of areas in their jurisdictions which could be affected by a breach of a high hazard dam and to transmit such review to the state engineer.[8]

All the back and forth about dam safety in Colorado resulted in the National Park Service Dam Safety Program releasing its first program guideline in 1983, and a strengthening of the Colorado State Engineer's Office by House Bill 1416. By 1984, a new organization, the National Association of Dam Safety Officials had been formed in Denver, which eventually became the Association of State Dam Safety Officials (ASDSO). An editorialist suggested "that when human life is in the balance, owners of suspect dams must be handed a 'fix or we'll drain it' order".[1]

# Removal of Dams

## Lost Lake

The Park Service began removals by breaching the Lost Lake Dam, also known as Glacier Dam #1, in the summer of 1987. They flew in equipment and a crew by helicopter and worked to breach the 11-foot-high rubble-masonry dam.[1] Although the Park Service had said the dam was in good condition following an inspection in 1982, and remote enough to not be a hazard, this mitigation was the easiest of the backcountry dams. The heirs of Nutter R. Lee sold the real estate and water rights to Lost Lake to the government in 1969, with final approval by the Larimer County Court on October 30, 1970. The final purchase cost of these water rights was $34,435. This lake drains "Lee's Fork" into the North Fork of the Big Thompson River.[9]

## Arbuckle Reservoirs

The Park Service worked from 1982 to 1987 negotiating with the City of Longmont to acquire land and water rights to the three Arbuckle dams in Wild Basin. The Park Service paid 1.9 million dollars to purchase these water storage rights[10], even though Longmont had the capacity to store their water in Ralph Price Reservoir, downstream on the North St. Vrain River. That allowed the Park Service to go forward with removing these dams, five years after the Lawn Lake flood.

## Pear Lake and Sandbeach Lake

In 1988, the removal of the Pear and Sandbeach Reservoirs non-concrete dams was completed. The U.S. Army provided a CH-47-D Chinook helicopter, and transported two bulldozers weighing 14,500 pounds to these sites[1]. Park Service crews removed all of the human-made portions of the dams, outlet structures, and outlet pipes. Crews worked to restore the lakes to the original natural settings and outlet streams.

Non-native fish were removed from the lakes, and native greenback trout were placed in both Sandbeach and Pear Lakes. The Park Service monitored natural revegetation of native species in the previously flooded shorelines and documented more than 50 native species following dam removal.[10]

## Bluebird Lake

In 1984, the Rocky Mountain Regional Office requested a Determination of Eligibility for the Bluebird Dam to be registered in the National Register of Historical Places as an intact example of an early 20th century concrete arch dam. It was cited as an important engineering achievement based on the logistical difficulty of high-altitude dam construction in the region. It was recorded as Eligible on December 19 1984, by the Keeper of the National Register. This record, E.O.11593, describes the dam and its construction details, documenting this structure prior to its removal in 1989 and 1990[10]. This "Section 106 Compliance" report was done as part of National Environmental Protection Act (NEPA) requirements. Despite the significant historic value of the early era dam construction, the safety and resource restoration values were considered more significant, which overrode the historic preservation designation.[11,12]

An environmental assessment (EA) was completed for NEPA, which considered four options for the dam. Dam removal was the final determination, while still considering all conceivable adverse impacts.[12]

In 1989 and 1990, the Bluebird Reservoir Concrete Dam was removed during two summer seasons. Because this is a high-altitude lake, the window for working without creating additional damage to the ecosystem was short each year. To reduce damage from work at the site, equipment and crews were flown in by helicopter. The original proposal for the concrete dam removal was blasting with dynamite, but because of a federally listed threatened species, the greenback cutthroat trout, the plan was altered. Because of the risk of concrete dust creating particulates, and altering the pH of the downstream creek, risking a fish kill, a different demolition technique was utilized.

*Bluebird Dam removal. See Schaeff machinery on dam-center right.*
*Photo courtesy, NPS (From Baker and McCormick 2012)*

A Schaeff Walking Excavator with independent articulating legs, and a long articulating arm with a hydraulic rock hammer at the end, was used for demolition. Demolition started on June 8, and the short season ended on September 22 1989. The Schaeff could work in the rubble as it was created, breaking up the dam, much like a Praying Mantis insect devours its prey. The exception being that concrete rubble and rebar could not be left in the environment, so five helicopters, including an Army Chinook and an Erickson Sikorsky S-64 Sky-crane, were used to remove the rubble from the sensitive high-altitude landscape. Five million pounds of concrete were removed from the dam site to landing zone seven miles away and then taken by dump trucks to an old borrow pit near Lily Lake on Highway 7. [1,12]

The concrete rubble stockpile waiting to be removed at the lake was held in an impermeable Hypalon rubber dike to prevent contamination of the lake. Downstream pH and turbidity were monitored daily, and reported out every five days during crew changes, to make sure environmental standards were maintained (J. Arnold, pers. comm., 2 February 2022). Absorbent barriers were positioned around equipment and fuel caches to avoid fuel spills into the environment that might impact the trout populations.[12]

Crews camped on platforms and walked on narrow pathways to prevent damage to the fragile ecosystem. They also used the old "dilapidated rock cabin" at the site as a kitchen.[12]

The majority of the dam demolition was done in 1989. In 1990, during the short summer season, the remaining dam and rubble material was removed.

While monitoring wildlife, it was found that Bighorn Sheep continued to use the landscape above the lake, but noise levels and the presence of people caused White-tailed Ptarmigan to leave the area during demolition periods. Wildlife surveys in 1992 found that most wildlife had returned close to pre-demolition levels.[9,12]

Natural revegetation for the lakeshores was monitored at all three Wild Basin dam removal sites, but willow cuttings from the dam sites were propagated and grown in the Park Service nursery and transplanted back at the de-dammed lake sites. Vegetation studies were done every year through 1993 and then every other year through 1998 to make sure plant diversity remained high (48 species at Bluebird).

The costs for dam removal for the Wild Basin Reservoirs include the 1.9 million dollars paid to obtain the water rights from the City of Longmont, and almost 2 million dollars in other costs.[9] Helicopter costs for 1989-1990 for the Bluebird Dam removal were $770,384.[10] Direct labor time for the project was 12,440 person hours. Weights for equipment flights and material removal totaled 5,559,300 pounds. No work injuries were reported during all three dam removal projects over three summers.[12]

To document the historical dam, a plaque was placed at the edge of the former dam site at Bluebird.

## Lawn Lake Dam

The Farmers Irrigation Ditch and Reservoir Company went bankrupt from the lawsuits and claims against them for the dam failure. Although the new owners of the Reorganized Farmers Ditch Company (RFDC), formerly FIDRC, requesting rebuilding the Lawn Lake dam, the National Park Service was strongly opposed. A new road would have needed to be constructed to rebuild the dam, and disruption of a threatened fish species, and dam safety, were among the reasons for opposing a new dam. In 1994, the NPS acquired the property rights to the dam, and in 2000, equipment was helicoptered to the dam site to restore the breach site. Two skid-steers and one mini-excavator redistributed 5300 cubic yards of dam remnants. The lake remained at the same level it was following the breach, since the dam no longer held back any water. The dam site was revegetated with native plants collected in the area.[1]

## Sprague Lake Dam/Dike

According to the NPS Dam Safety Office, Sprague Lake Dam is considered a Low Hazard potential dam/dike. In 2011, the dam was damaged when several large trees fell during a windstorm. It was repaired in 2012.[1] Additional work on the Sprague Lake dam was done in 2017. A seepage control berm was added, spillway improvements were made, portions of the dam crest were elevated, and construction of an inlet control structure were done. Old tree stumps were removed, as well as other potential hazard trees on the dam. The lakeshore had riprap added to prevent dam erosion, and excess flows estimated greater than a 10-year flood were engineered to be diverted over the Sprague Lake Road into Glacier Creek. Finally, a new pedestrian bridge and trail improvements were made at the spillway.[13]

## Copeland Lake

The Copeland Lake level was lowered four feet, and now is permitted to hold up to 100-acre feet of "augmentation" water for use by the City

of Longmont through the St. Vrain and Left Hand Water Conservancy
District (SVLHWCD).

In a lease agreement between the SVLHWCD and the National Park
Service in December 2007, and renewed annually, Copeland Reservoir
is managed by the NPS as property consistent with the purposes of
Rocky Mountain National Park, as part of Wild Basin. Water from
Copeland Lake and the North St. Vrain flow into Button Rock Dam/
Ralph Price Reservoir, and then to a Diversion structure near Lyons
before flowing to Longmont.[14]

During the 2013 massive Front Range storm event, Copeland Lake
and the Wild Basin Trailhead were posted with this sign on 9/22/2013.[15]
*TEMPORARY CLOSURE: Due to flooding and structural damage this
area is closed until further notice.*

The 2013 flood events damaged almost all stream beds, challenged
all existing dam structures, and caused road washouts on Colorado
State Highways 34, 36, and 7, as well as causing massive flooding in
Glen Haven, Drake, the Big Thompson Canyon, and Lyons, Colorado.

## Lily Lake: A National Park Service Success Story

In May of 1951, the Lily Lake Dam failed. Rather than a rainstorm-caused
event, the investigation found that the failure was a result of high winds
causing wave action over the dam. "Terrific lashing waves pounded a
hole through the earth dike sometime prior to 0500 [AM]…releasing
lake water". A witness saw waves 2 to 3 feet high on the lake. The investigation states that about 75 acre-feet of water was released, which was
the stated capacity of the lake. Another eyewitness stated that the water
was overflowing State Highway No. 7, and that within 10 minutes *"the
paved surface was lifted and was carried down into the canyon, trees and
boulders crashing with it"*. There were no injuries or loss of lives, but
property and road damage were estimated at $10,000.[16]

In 1979, I had an offer to live in a cabin at Lily Lake from the care-
taker for the Charles Mace family (Baldpate Inn), but I felt that it would

*Breach in State Highway 7 just downstream from the dam¹. Photo, CDOT*

be too far away from Estes Park to be on call for the ambulance service. I regret the missed beauty and experience from not living in that location that year.

Lily Lake was private property until 1989 when it was purchased by the Conservation Fund to prevent development. Lily Lake became part of Rocky Mountain National Park in 1991. The dam was rated by the Bureau of Reclamation as a High Hazard Dam in 1992. The National Park Service (NPS) began looking at improvements for parking and a visitor center/trailhead in 2009, and requested public comments on alternatives including the removal or repair of the dam by February 2012.[17]

The decision was made to repair the dam and make improvements for visitation there and on the east side of Highway 7. Mark Baker wrote, *"during the time of public review, the park discovered that there were water rights associated with the purchase of the dam and land during the 1990's. Legal staff determined that the NPS could not divest itself of the dam. The park made the decision to repair the dam."*[18]

*Lily Lake Dam following dam repair – circa December 2012. Photo courtesy of the Bureau of Reclamation.*

This appears to be a very timely decision. With a new classification hazard report in 2011, and a new 2012 risk chart, the risks of Lily Lake dam failing were the highest in the NPS inventory of dams. With an "EXPEDITED" label placed on all correspondence by the NPS Dam Safety Officer regarding this dam, the process proceeded quickly to the repair phase of Lily Lake Dam. The plan was for construction between July and December 2012. Construction actually began in August 2012, and after a delay from a snowstorm in October, was *"substantially completed"* by December 2012.[18]

The NPS had in place an Early Warning System, and an Emergency Action Plan (EAP) Exercise had been held with 21 agencies on April 5, 2012.

Nine months after the Lily Lake Dam repair was completed, the Front Range of Colorado experienced another "thousand-year flood event" (NOAA). Beginning on September 9, 2013, a large and slow-moving storm stalled over the mountains from west of Colorado Springs, north to the Colorado-Wyoming border. It was estimated that from September 9th to September 15th, anywhere from nine to seventeen inches of rain fell in this broad area.

For Lily Lake, the NPS Dam Safety Officer received text messages transmitted through the Early Warning System to the National Monitoring Center (BIA-NMC), that the Lily Lake precipitation accumulation was greater than 2.4 inches in the past 6 hours (25-year rainfall event), followed by a message five minutes later that 3.5 inches of rain had fallen in the last 24 hours. An hour later, a message stated that 4.3 inches of rain had fallen in 24 hours, which corresponded to a 100-year rainfall event. Personnel from the Park Service and other agencies responded to monitor the dam and assess damage.

*Lily Lake 2020*

There was some damage to the "downstream toe outfall and the inspection well area" of the dam. The gravel and soil were washed from the base of the spillway, and large boulders were washed into the spill-way from the upper slopes.[18]

Fish Creek itself washed out and caused considerable damage to sewer, water and electric lines, pedestrian and bicycle paths, Fish Creek Road and crossroad intersections. However, this damage was due to the heavy rains and runoff, not a failure of the Lily Lake Dam. Originally estimated damage of around five million dollars for repairs of the damage from Fish Creek flooding, ended up exceeding twenty million dollars with added projects and a consolidation of resources.[19]

The National Park Service did everything right in this case regarding the Lily Lake Dam.

There are nice walking trails around the lake, and on Lily Mountain, and fishing enhances this enjoyable park experience. Fall colors from the many Aspen trees around the lake and adjacent mountains are always worth a trip here in September. An experimental Greenback Cutthroat trout population was introduced to this lake in 1990. Although it is not a breeding, sustainable population, annual stocking, and an oxygenation system used in the fall and winter provide support for this population in this relatively shallow lake.

The Rocky Mountain National Park website has the following blurb on Lily Lake.[20]

Lily Lake is easily accessible and a great place to look for paedo-morphic tiger salamanders in Rocky Mountain National Park.

*This paedomorphic tiger salamander* (Ambystoma tigrinum) *is sexually mature, but retains some larval characteristics including its obvious frill of gills. Photo courtesy of the Greater Yellowstone Network Vital Sign Monitoring and USGS.*

## Cascade Lake

The Cascade Dam was lost secondarily in the Lawn Lake Flood. It caused two of the three deaths due to its failure, when it flooded Aspenglen Campground. The Town of Estes Park lost its Stanley Powerplant which had operated continuously for 72 years. The diversion dam and penstock at Cascade Lake were repaired in 1988, to supply the Fall River Treatment Plant with water.

In 1991, Estes Park filed an application for water rights at the Cascade Diversion Structure. In March of 1992, the Water Court adjudicated a direct flow water right of 1.55 CFS for the Cascade Diversion and a 3.45 CFS conditional water right for the Estes Park Cascade Diversion Enlargement.[21]

The Town of Estes Park filed a Declaration of Intention to restore the hydropower generating station with the Federal Energy Regulatory Commission (FERC) on March 9, 1994. Although the Town claimed that FERC had no jurisdiction over the generating station, the Park Service held the view that FERC needed to approve rebuilding the dam, which was on federal Park Service land, even if the power plant was on private

*Care-taker's Cottage and Penstocks from the Stanley Powerplant.*

*Cascade Diversion Dam and Penstock rebuilt in 1988.*

*Flood eroded hillside adjacent to Cascade Lake and Cottages.*

*Some of the restored powerplant workings in the museum, the clock on the right wall stopped when the water level reached that height on the wall.*

lands[21]. Opposition by the National Park Service and conservation groups for not rebuilding the Cascade Lake dam won out.

A decision was made not to restore the Cascade Lake Dam and the generating station, and a compromise was reached. In 1997, the Town of Estes Park and Rocky Mountain National Park, and the utility provider agreed not to rebuild the plant, in exchange for supplying 500,000 kilo-watts of power to the town.[21]

In May of 2011, the Town of Estes Park filed an application to discontinue the 3.45 CFS conditional water right, which would be then terminated by the Water Court. They kept in place their absolute right of 1.55 CFS for the Cascade Diversion.[23]

The historic Stanley Fall River Hydroplant was restored as an interpretive museum using $400,000 in funds provided by the Colorado State Historical Fund[3], and placed on the National Register of Historic Places. It was dedicated on the 20th anniversary of the Lawn Lake Flood on July 15, 2002. It was painted with bright yellow paint that was also the original Stanley Hotel color.[24] The powerplant Operator's Cottage is also part of the restored property.

Adjacent to Cascade Lake, the privately held Cascade Cottages and the 43 acres of property with the cabins, were purchased by the Rocky Mountain Conservancy through the Trust for Public Lands (TPL). TPL transferred this property to Rocky Mountain National Park ownership in March of 2017.[25] The Cascade Cottages property was entered in the National Historical Places Register in 2020.

# CHAPTER EIGHT

# The 2013 Front Range Floods of Colorado

In September 2013, a late season storm stalled on the Front Range of Colorado, dumping up to seventeen inches of rain over a period of a week to ten days. This storm was similar but much larger than the stalled thunderstorm that built over Dry and Devil's Gulch in Estes Park that caused the 1976 Big Thompson Flood. This large regional storm caused flooding from the south of the Colorado Springs area to areas north of Fort Collins. The long duration of the storm and volume of water created saturated soils, resulting in high runoff to creeks and rivers, and filling dams to capacity, or above their spillway ability. This resulted in nine low-hazard dams failing and extensive flooding throughout the entire region.[1]

The flooding ran downstream from the mountains to the plains and included six major rivers and their tributaries.[2] More than sixteen cities and towns were flooded or had some flooding. Manitou Springs, Nederland, Jamestown, Boulder, Lyons, Estes Park, Pinewood Springs, Glenhaven, Drake, Evans, Greeley, Fort Collins, Bellvue, Loveland, and Longmont were hardest hit among them. Flooding of the South Platte River continued east all the way to Fort Morgan. Fox and West creeks above Glenhaven, the North Fork BT, and the Big Thompson River between Estes Park and Loveland flooded catastrophically again. Highway 34 through the Big Thompson Canyon was washed out on September 11. Many residents had to go back up the canyon to Estes Park for shelter, or those in the lower canyon, to Loveland. Many others were stranded and had to be rescued again, just 37 years after the original Big Thompson

Flood. Fish Creek, below the Lily Lake dam, which drains five miles through Estes Park, ran at an estimated flow of 6,900 cfs (see Table 1), resulting in extensive flooding, taking out nine road crossings and damaged the sewer lines, dumping raw sewage into the Big Thompson River.

The three main highways to the east from Estes Park were washed out, limiting eastern travel via Highway 7 through Nederland and Boulder. Highway 34 over Trail Ridge Road was the only western route available.

Steven Yochum and Daniel Moore[3] examined peak stream flow estimates from the 2013 Front Range Floods, and I have summarized some of their numbers here in Table 1, especially for areas in proximity to Rocky Mountain National Park, to compare with estimates from other floods.

Table 1: **Selected Peak Flow Estimates from streams and rivers, 2013—Colorado Front Range**

| Location | Average Peak Discharge (cfs) | Multiples of 100-year flow |
|---|---|---|
| Little Thompson River at Pinewood Springs | 14,600 | ~ 3 times |
| Fish Creek upstream of Lake Estes | 6,900 | ~ 5 times |
| Fall River upstream of Estes Park | 3,800 | ~ 2 times |
| West Creek upstream of Glen Haven | 11,000 | ~ 4 times |
| Fox Creek upstream of Glen Haven | 3,500 | ~ 3 times |
| North Fork BT River upstream of Glen Haven | 1,700 | 50-100 year flow |
| North Fork BT River upstream of Drake | 18,400 | ~ 2.5 times |

*Modified from Table 1. Yochum & Moore 2013. Locations selected by proximity to Rocky Mountain National Park

The combined flows from West Creek, Fox Creek, and the North Fork of the Big Thompson above the town of Glen Haven were estimated at 16,200 cubic feet per second. Upstream of Drake, the flow of the North Fork of the Big Thompson was estimated at 18,400 cfs.[3]

These flows in 2013 were devastating to the canyons and roads that enter Estes Park from the east.

During the Big Thompson Flood of 1976, estimates for the North Fork of the BT at Drake were 8,700 cfs, and the main Big Thompson canyon at 28,000 cfs. At the far end of the canyon past The Narrows, the flow was estimated at 31,000 cfs.

Estimated flows from the Lawn Lake Flood in 1982 were a peak flow of 18,000 cfs from the breach of Lawn Lake Dam, and 16,000 cfs from the breach of Cascade Dam. Water flows entering Estes Park during this flood were estimated at peak flow of 6,000 cfs.[4]

Nine low hazard dams overtopped or failed in 2013, including five dams at Big Elk Meadows, Carriage Hills #2 dam in Estes Park, Havana Street Dam in Commerce City (Denver), and two Emerald Valley dams in Colorado Springs. Button Rock Dam on the North Saint Vrain Creek west of Lyons overtopped the spillway without breaching, with approximately 10,000 cfs flowing into the Town of Lyons, cutting off portions of

*"Mass Movement," West Side of Twin Sister's Mountain, RMNP.*

the town and causing devastating damage. A low hazard dam is one for which minor damage to property is expected with no anticipated loss of life.[1] Two "mass movements" or debris flows[5], described in Yochum and Moore (2013)[3], were landslides/mudslides that occurred on Twin Sister's Peak, one on the west side and one on the east side. The scars are easily visible, the slide on the west side required a rebuilding and re-routing of the Twin Sister's Trail where it crosses that slide. The eastern Twin Sister's slide scar can be seen from the eastern plains when there are contrasting conditions to highlight it against the green hue of the forest (See Neníisótoyóú'u photograph in Chapter 2). Another debris flow occurred on the east face of Mount Meeker, where flooding from Cabin Creek formed an alluvial fan down the hillside, that damaged the Camp St. Malo property and Highway 7.

To add insult to injury, within 12 days of the end of the storm, while people were still marooned, displaced, or unable to travel due to damaged roads, the Congress of the United States shut down the Federal Government for 13 days beginning on October 1. This was due to their inability to pass a budget for Fiscal Year 2014. This included shutting down Rocky Mountain National Park, including the only roadway open to Estes Park from the west. We were fortunate that Colorado Governor John Hickenlooper provided state funds and resources to reopen the National Park and fund highway repairs during this critical shutdown period.

Eight people died in this extensive flooding. Jamestown was isolated and flooded with seventeen inches of rain in three days, resulting in one death. Jamestown damages exceeded $22 million dollars.[2] Boulder Creek coming down Boulder Canyon from Nederland, caused three deaths, flooded 900 square miles, destroyed 262 homes, and damaged 300 other homes.[2] Fountain Creek, near Colorado Springs, flooded El Paso County, which had two deaths. Nineteen inches of rain was reported near there in Fort Carson. One death was reported in Cedar Creek and one death in Clear Creek Canyon.

In Larimer County, 1500 homes and 200 businesses were destroyed, with many more damaged. Boulder County had 262 homes destroyed and 300 damaged. In Weld County, the South Platte River damaged 3000 homes, 350 businesses, 2377 crop fields, and 1900 gas wells. To the east, seventy-three miles of roads were damaged in Logan County by the South Platte River as well. A total of one to two billion dollars in damages was estimated for the Front Range and Eastern Plains.[2]

Perhaps, because the rain was more gradual and a long-term event, modern technology, early warning systems and better communication systems resulted in far fewer deaths statewide than the Big Thompson Flood of 1976.

John Batka, a Dam Safety Engineer with the State of Colorado, summarized his paper with this statement: "*The 2013 flooding event was the 7th historic flood in Colorado since 1902, suggesting a frequency of about once every 16 years. This suggests that every dam safety engineer planning on a career of 20 years or more can reasonably expect to be involved in such an event and therefore should prepare for it. It is our hope that other state dam safety programs will take the lessons and information from our shortcomings and successes and do their best to protect the safety of their citizens the next time a historic flood event occurs*".[1]

Since the rest of us are also likely to experience at least one flood event in our lifetime, use caution when traveling through canyons and in the high country, and remember to "Climb to Safety in Case of a Flash Flood!"

# CHAPTER NINE

# Return to the De-dammed Lakes

I n 2014, I decided to hike into as many of the de-dammed lakes in Rocky Mountain National Park as reasonable in one season. I hiked into Bluebird Lake, Sandbeach Lake, Pear Lake and finally, Lawn Lake. My goal was to see the natural lakes after they had been rehabilitated following removal of the dams and look at the water along the way that makes the Park so special to the hiker.

## Bluebird Lake

My first hike was July 19 to Bluebird Lake, with my 18-year-old grand- daughter as a hiking companion. Highway 7 from Lyons to Wild Basin had been temporarily repaired since the massive floods of 2013, but the scoured and damaged river course of the South Saint Vrain River was still highly visible along the way. Large out of place boulders, and new creek courses and gravel bars demonstrated that rivers

*North Saint Vrain River, Wild Basin*

only hold temporary courses, and what lies along the banks may be temporary as well. Trees that had held their place for hundreds of years were down, and houses that had shared space with the creek for many years were damaged or destroyed.

Wild Basin Trailhead sits about two miles from the highway and follows the North Saint Vrain River to the parking lot. It is the water that makes this part of the Park so special, from the drive in, to the trail that courses mostly along the creek. I can close my eyes as I walk here,

*Damage from the 2013 floods, North St. Vrain River.*

*Upper Copeland Falls, Wild Basin*

and know where I am on this trail by the sounds and smells of where I am. Hunter's Creek first crosses the trail near the trailhead, and then the little but beautiful Copeland Falls and Upper Copeland Falls appear. Short stretches of trail avoid the river, but water is never far away.

We took a break at the old CCC Bridge over the North St. Vrain River, then went back down trail a few yards to the cutoff at the Pine Ridge campsites, which is the unimproved "Fishermen's Trail". Since the bridge at Ouzel Falls had been taken out by the 2013 floods, we needed to head this way to get to the Ouzel-Bluebird Trail, due to the trail closure. I later wrote an article about the new bridge at Ouzel Falls after it was replaced, in 2016.[1]

*The view upstream from the CCC Bridge-Wild Basin.*

We would have to backtrack down trail from the North St. Vrain Bridge, where the fisherman's trail meets the main Thunder Lake Trail. One-half mile down trail from this bridge is the Ouzel-Bluebird Trail Junction. At this junction is the old snow-course measuring stick that had been there for likely over 100 years, and may have been part of Enos Mills' snow-course measurements in this area in the early 1900's as the "State Snow Observer of Colorado".[2] (See first photo this page.)

As we climbed up Ouzel Ridge, the views of Long's and Meeker poke out of the new forest regrowth from the Ouzel Fire in 1978.

My grand-daughter Stephanie painted this scene from one of our photos from this trip, it sits in our living room along with K. Z. Berquist's beautiful Ouzel Falls painting.

The last time I had been on this ridge was in 1979, when the ground was completely denuded with the dusty exposed trail passing through the dead skeleton

*Above: Long's (left) and Meeker Peaks- From Ouzel Ridge.*
*Top: Snow-course pole by Ouzel-Bluebird Junction.*

trees. It was refreshing to see the new luxuriant growth after working in the devastation from the fire all those years ago. From this ridge while mopping up the Ouzel Fire for over a month, we would fill our backpack pumps in Ouzel Creek and hike up Mahana Peak to work on the active fire.

*top: Trail along Ouzel Ridge after fire, October 1978.*

*left: The scars amongst the beauty of regrowth.*

*above: Ouzel-Bluebird Junction, Mahana Peak to the left.*

After this trail junction, I told my grand-daughter that the last time I had seen a moose in Wild Basin was at Chickadee Pond in the

1970's. Right after I made that statement, two bull moose walked out of the meadow adjacent to the pond, also see the photo in chapter two. We spent considerable time photographing these two moose, while protected by a nice rock field between the trail, us, and the moose.

As we continued up the unimproved trail to Bluebird Lake, water continued to be a big part of the landscape.

*Bull moose near Chickadee Pond*

Far away falls cascading down the cliffs were stunning as were the closer water features. Snow was still present nearby the trail, and where the snow had receded, the yellow Glacier Lilies filled in the wet soil.

This was the more difficult part of the hike, puffing away up a wet, steep trail, but the scenery was really what took your breath away. I had forgotten what made this long trek worthwhile, and knew that from where the water came, the lake had to be close.

*Falls cascading and trickling down through the snow.*

*Yellow Glacier Lilies*

*The Bluebird Lake trail still snowy in late July.*

A bit more of a trek uphill through snow and up a rocky path and we were there. The last time I was here in 1979, building a footbridge, the big, ugly dam interfered with the hope of a pristine view. It did take away from the National Park wilderness experience, except when you consider the historical background of the Arbuckle Reservoirs that preceded the Park.

*above: Restored Bluebird Lake, with Ouzel Peak in the background.*

*right: The wildflowers added to the beauty of it all-King's Crown.*

*Krummholz and landscape of Bluebird Lake.*

*Very little scarring, essentially no evidence of a dam seen.*

When the clouds came in and sleeted on us during a late lunch, we hid in the Krummholz (see photo page 135). There, I found the only evidence of the former dam site, a half-buried old metal bucket. There is a commemorative plaque placed at the former dam site, but we didn't see it in our haste to leave the area before more thunderstorms came in.

The hike out was brutal, I was wearing worn out 35-year-old boots, my grand-daughter, running shoes. The downhill jamming of our toes and the exhaustion of the long hike to the trailhead reminded us that this was not an easy quick hike for the reward of the beautiful view of the restored nature of the de-dammed Bluebird Lake. Congratulations to the Park Service for such a high-quality result.

## Sandbeach Lake

My second hike to the de-dammed lakes was on August 9, 2014. The trailhead for Sandbeach Lake is just off Highway 7 as

*Above: Velvet-antlered Mule Deer; top: Sandbeach Lake trail sign.*

you enter the Wild Basin area, where the entrance station is, just before Copeland Lake. The bear-proof containers at the trailhead remind hikers not to leave food in their cars. The trail for Sandbeach starts right up the moraine, switch-backing up the hillside from the valley. At 6:14 in the morning, the cool start helps to motivate me to get up this steep hill. The signs point the way, and a mule deer in velvet is traversing the hillside as well.

Hiking along the Copeland Moraine, you can see the burn scar from the Ouzel Fire on Meadow Mountain, the 60-foot-high flames heading for the Town of Allenspark, prompted the Park Service to call in the Boise Interagency Fire Center (BIFC) in 1978, now renamed the National Interagency Fire Center (NIFC).

Hiking up the Copeland Moraine can be the less exciting part of this hike, but the views of the Wild Basin valley where the North Saint Vrain River meanders, show the complexity of the riparian landscape. In the fall, the ridge on the opposite side of the valley is resplendent in changing colors, hence the name Fantasy Ridge.

Here I saw chickadees working the trees and a hummingbird resting on a branch. The first water on the trail comes with Camper's Creek, which drains off Horsetooth and Lookout Mountain. There are campsites

*At the top of Copeland Moraine, the trail junction to Meeker Park.*

nearby, Hole-In-The-Wall, and Camper's Creek sites, but it is yet early in the hike, 2.3 miles from the trailhead.[3]

The scenery is beautiful in the forest, Pinedrops, lichen-covered rocks, wildflowers, and mushrooms were springing through the soil. One mile further, and Hunter's Creek appears.

*The Ouzel Burn Scar on Meadow Mountain.*

*Wild Basin valley, North Saint Vrain River.*

*Camper's Creek.*

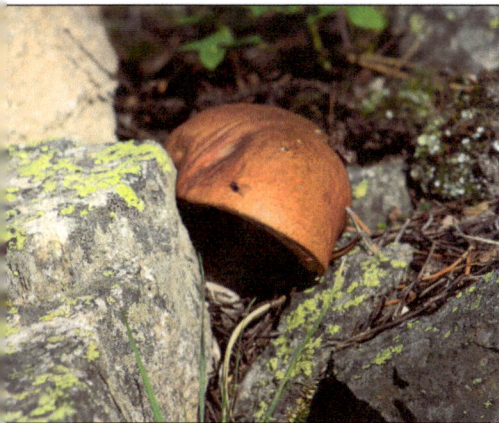

*Mushroom cap emerging between the rocks.*

*Lichen on Rocks-Sandbeach Trail.*

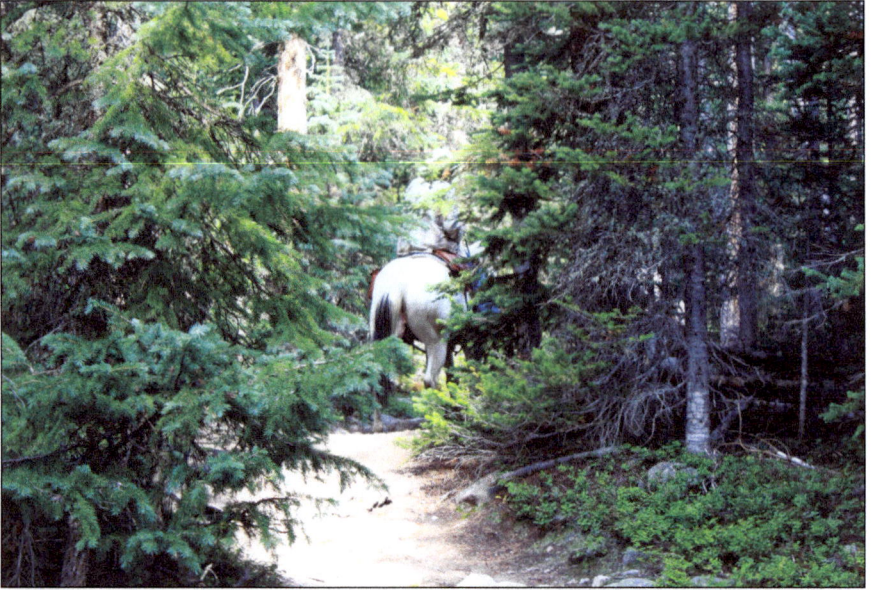

The trail is shared by hikers and horses.

Hunter's Creek, no fishing, for protecting Greenback trout.

After Hunter's Creek, the trail becomes steep, I paused to let a group of horseback riders pass by. The steep section is tough on hikers and horses alike because of the rocks.

One more mile of pretty steep climbing to be rewarded at the top of the hill. Past the hitching post and down through the trees, the lake emerges.

*top left:* Amanita *mushrooms; top right: Log checks and Rock bars prevent down trail erosion; above: Sandbeach Lake emerges through the trees.*

*Sandbeach Lake-Copeland Mountain is in the center right.*

*Looking north is Mount Meeker, Long's Peak is far left-flat summit.*

With the restored, de-dammed Sandbeach Lake, the sandy beaches are again prominent, and a rocky peninsula is no longer submerged.

Water trickles in from the west off Mount Orton, and Sandbeach Creek flows south out of the lake before heading east down to the valley below. Hiking around the lake, many young trees have sprouted around the lake, amongst the rocks. The older, more heavily forested side of the lake comes down from the west side of Mount Orton.

Where there is water trickling in, the wildflowers are abundant, as well as *Amanita* mushrooms, growing in the thick soil.

On the outlet side of the lake, there is no evidence of the former dam, and the outlet is a nice trickling stream.

The lake and the views are spectacular, and the sandy beaches are more exposed since the lake was restored to its natural levels. This hike,

*Sandbeach outlet, Pagoda Mountain, Long's and Meeker Peaks.*

*The classic sandy beaches for which the lake is named.*

although there were steep sections of trail, is only 4.5 miles in, and well worth the hike for the scenery at the end of the hike. As is classic with many of the lake hikes in the Park, there is plenty of water along the way to enhance the beauty of the route. On the hike out, I soaked my sore feet in Hunter's Creek, and while I sat just off trail in the woods, a family of Blue Grouse ambled by right next to me.

## Pear Lake

My third hike to a de-dammed lake was to Pear Lake in Wild Basin on August 23, 2014. On the drive-up Colorado Highway 7 from Lyons, the southern view of these four peaks was stunning.

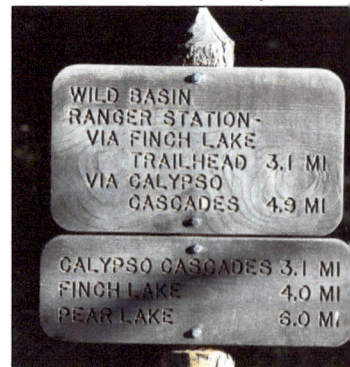

*The destination, Pear Lake, 6 miles*

At the Allenspark Trailhead at 7:22am, it would be cool enough for this first stretch of two miles to barely break a sweat.

The sign outlines the destination, Pear Lake 6 miles. This is what I consider the back way, unfortunately it is the drier choice of a trail, at least for the first few miles.

*top: Chief's Head, Pagoda, Neníisótoyóu'u (Long's, Meeker); above: Allenspark Trail.*

On the way, I looked for an old cabin I found while fighting the Ouzel Fire but did not find it today. This trail follows the old road up to the former reservoir, along the side of Meadow Mountain. Except for a few flat sections, this trail is fairly steep as it climbs to the Finch Lake/ Calypso Cascade junction at the top of the hill.

A few very small creeks cross the trail, and there is a junction down the hill to the "Finch Lake Cut-off Trail", which is a pretty hike.

The top of the hill brought great views, a fresh snow on the mountain tops and clouds add to the beauty. Dark-eyed Juncos and

*top left: Golden-mantled Ground Squirrel; top right: Everywhere there is art, you can almost see a face in this burl; above: Fresh snow on the mountain top.*

Golden-mantled ground squirrels hung out in the cool morning, foraging while it is still quiet.

The hike becomes more interesting, there are meadows slanted across the hillsides with wildflowers, and water trickling down across the trail. A Clark's Nutcracker poses in a treetop.

*Clark's Nutcracker*

Across a few bog bridges and around a few corners, opens up the view through the trees of Finch Lake, about four miles in. Finch Lake is a very shallow, relatively warm lake, that I learned had leeches, usually discovered when swimming. The outlet creek that flows into Cony Creek is warm, contrasting with the ice-cold water coming down Cony from Pear Lake. We were camped out here for 40 days in my first summer with the Park Service, when it mostly rained all day, while we cut trees to widen the trail.

*Finch Lake*

Very quickly past Finch Lake, the trail crosses Cony Creek. Up past some meadows, and a pond with water lilies, the trail steepens as it heads up the two remaining miles to Pear Lake.

I half expected to see moose here and did on the way back down the trail.

Hiking through meadows and wildflowers, across old bog bridges, and a less improved trail as I climbed higher, I passed the campsites at Pear Creek, which flows out of Pear Lake. These last two miles from Finch Lake are more of a wilderness trail than the wide horse and pedestrian trail up to Finch Lake. There is plenty of water and wet areas along the trail, which support a variety of wild-flowers along the way. One of my favorite succulents is Rose Crown, downgraded in color from the dark red of its cousin King's Crown, to the pink-rosy color that defines its name.

*top right: Rose Crown; above: Pond with Water Lilies.*

Up the last stretch, the sign for the Pear Creek campsites and a short uphill push. Bog bridges and wildflowers in the moist meadows tell us that water is near.

At last, the lake, the old high-water line can be seen in the rocks. It was a little bit blustery. You can see white caps on the water.

*top: The final stretch, Copeland Mountain is center right; above: Pear Lake outlet, note the former high-water line.*

From here it is hard to see the fat pear shape of the natural lake, named by Joe Mills. Elk Tooth, Ogallala Peak, and Copeland mountains form the cirque that hold Hutcheson, Cony and Pear lakes on this southeast border of the Park. Elk Tooth is outside the Park boundary. As you walk around the lake, you can see the former high-water mark of the reservoir, but since it hadn't held that volume of water for quite some time, the bathtub ring effect is minimally seen.

Pear Lake isn't quite as spectacular as Bluebird Lake but holds plenty of natural beauty and nice alpine views. For the more adventuresome, hiking to Cony and Hutcheson Lakes will give you more of a cross-country wilderness experience and spectacular views of this high cirque on the Continental Divide.

*Restored Pear Lake, Copeland Mountain is to the right.*

*Old water line of lake seen with bare rock under lichen pattern.*

*Time for lunch and a nap.*

The trip down is "just a walk in the park", water adjacent to the trail is always a nice feature.

As I mentioned before, I half expected to see moose along this trail, and on the way down, I found this calf and her mother just off trail, as I headed for Cony Creek.

Mom was nearby, so I made sure I kept my distance and a jumble of downed logs between us for protection. Cony Creek is always a welcome crossing before Finch Lake. If you follow the fisherman's trail down Cony Creek, it

*above: Water and wildflowers, always a nice mix on the trail; top:Calf moose foraging in the luxuriant meadow*

takes you to Calypso Cascades on the main Wild Basin Trail. This was always our quick shortcut out, when we were camped out at Finch Lake. Unfortunately, it is not a good trail to take back uphill, so the hike back in was on the main trail. Finch Lake comes quickly going back down past Cony Creek, and then the trek through the forest and across the meadows of Meadow Mountain before coming to the trail junction.

This trail back out is not nearly as strenuous as the trek out of Bluebird Lake, so by the time I got to the Allenspark Trailhead, I was not nearly as exhausted. That completed my third hike to the former dammed Arbuckle Reservoirs in Wild Basin.

*Cony Creek above Finch Lake*

## Lawn Lake

The next restored lake I hiked to was Lawn Lake. The Lawn Lake Dam failed on July 15, 1982.

*Lawn Lake Trailhead*

I arrived at the trailhead around 6:30am on September 6, 2014. It was a foggy, misty morning that guaranteed a cool, wet hike.

The Lawn Lake Trailhead is where Steve Gillette, while picking up trash, used an emergency phone to call to the Park Service when he heard crashing noises coming down the mountainside from the flood on July 15, 1982. Not much had changed, except an expansion of the trailhead parking lot and bear- protected trash containers.

As I headed up the trail on Bighorn Mountain, I could see remnants of the alluvial fan down the hillside. The new dead trees were likely a result of last year's flood.

As I hiked, there was much evidence of the flood damage to the Roaring River drainage and to the trail. The 2013 floods have also con-

*Alluvial Fan-Roaring River-Bighorn Mountain*

*The Roaring River Canyon created in 1982.*

*Roaring River and adjacent trail.*

tributed to the visible erosion in the relatively new canyon, but the young tree growth in the cut demonstrates the renewal since the 1982 Lawn Lake Flood. Without solid ground cover, these bare slopes easily erode with each large rain event. Much of the Lawn Lake trail was adjacent to this newly created canyon edge and was washed out during the flood. This old trail was lined with many beautiful lichen-covered rocks and would have to be relocated, losing much of the character of the old path.

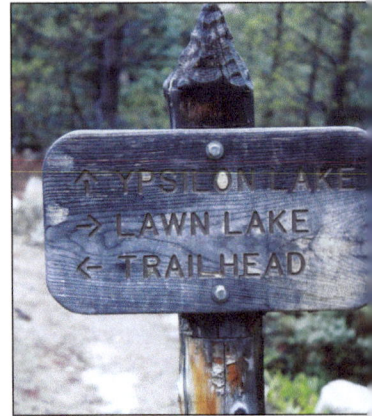

*Ypsilon-Lawn Lake Trail Junction*

After traversing up the first section of trail on Bighorn Mountain, 1.4 miles in is the trail junction to Ypsilon and Spectacle Lakes. Where the Roaring River cut through here during the flood, campsites at Roaring River were flooded, resulting in one death, and the remaining campers were stranded from hiking out and had to be rescued. At this junction, there are still 4.8 miles to go to reach Lawn Lake.

*Trees washed up from the Lawn Lake Flood along Roaring River.*

Hiking on toward Lawn Lake, there are many areas where the uprooted trees were piled up from the flood cutting its path along the river. While the evidence of the flood is everywhere close to the river, there are sections where the trail is serene, and many wildflower patches are found. This section of trail is relatively flat for a while before more switchbacks come to climb up Bighorn Mountain.

Along the path of the river, the raw exposed soil and deep ravine seem surreal while the late summer flow is little more than small stream in a big cut. Had the narratives of the flood not been so intensely horrifying, it would be hard to comprehend the rush of a huge wall of water down the mountain.

*Lichen rocks line the trail.*

It is so peaceful here now, the forest is quiet, except for an occasional bird call or the sound of water nearby. The humidity is high, the trees are dripping with water, and I am well soaked hiking in the cool of the morning. I have seen no other hikers on the trail this morning so far.

After a nice stretch of hiking, I am forced on a trail detour, reminding me again of the flood's force of water in this old well-established forest. These trees have worked hard to establish roots in an already rocky soil. I knew that the funds would eventually become available to rebuild these trails, and hire trail crew such as I was, to work in these spectacular places. I used to tell people that "I got paid to hike fast", as well as be in great shape and work hard in the backcountry, which was just about the best job one could have in the National Park Service.

Despite the damage, the hike remains beautiful. Glimpses of an almost ghostlike Mummy Mountain appear in the distance.

*See the remaining trail in the upper center.*

*Lichen covered rocks along the trail.*

Along with the rocks, a couple of small creeks trickle across the trail.

A short way below Lawn Lake, about 30 minutes by my photo log, I came across a sandy

*Roaring River washed out down to the bedrock.*

flat, on closer examination, I found part of a large metal pipe. I'm not sure if this was part of the dam structure, although it was not the outlet pipe seen in the post-failure photographs of the breach, nor the valve itself. The sand was deposited during the flood, as well as the pipe debris.

I looked for other metal debris in this area but did not find anything but the pipe remnant.

*Gray Jay in treetop*

*Sand deposit from the flood*

*Piece of metal pipe, center left, found downstream from Lawn Lake.*

Along the trail there was more sand and larger rock deposits from the flood. The wide, scoured Roaring River channel was very visible between the trees looking toward Lawn Lake. The river, now just a gentle flow out of Lawn Lake, is little more than a gentle flowing trickle at the bottom of this floodplain.

Shortly, I came to the Potts Puddle cutoff trail, about six-tenths of a mile below the lake. It was on this trail to Potts Puddle in 1980 where the trail crew I was working with, was drilling for blasting to improve the trail.

*Roaring River channel below Lawn Lake.*

*left: Potts Puddle (Black Canyon) Trail Junction; right: Last night's precipitation, visible in the shelter of this rock.*

*Roaring River, bottom left, flood channel above.*

On this last stretch before the lake, I passed a gentleman with a heavy pack making his way down trail, apparently everyone at Lawn Lake last night got drenched in a good first winter-like, rain-snow mix. At 10,987 feet high, winter-like weather is not unexpected in early September.

As I approached the lake, the breach from below is obvious, despite the rehabilitation of the site. The gap is still there, but it has been rear-ranged back into a more natural state, with a small falls cascading below, and willows and other natural vegetation growing in the breach. The clear blue skies and the views of the mountains are spectacular. It is easy to see what a beautiful subalpine setting this is, for just a moderate hike of a little more than six miles. I can't wait to see the restored lake.

*Below the breach, the last stretch to the lake.*

*A closer look at the breach from below Lawn Lake.*

*A small falls trickles down from the breach at Lawn Lake.*

*A subalpine meadow (lawn) and Mummy Mountain.*

*Lawn Lake–see former high-water mark on rocks–center, right.*

Hiking around the outlet end of the lake, I can easily see the high-water line on the rocks around the lake, but the vegetation has grown back in around the edges. Remembering the lake as it was before the flood, the straight line of the dam certainly wasn't natural looking, but the site was still beautiful when the lake was

*Yellow-belly marmots*

full. I think that the roughly 16-acre natural lake has more pristine beauty than the full 48-acre reservoir. Either way, this location is well worth the hike. Beyond the lake to the north lies an unnamed pond, Little Crystal Lake (3.7-acres), just a little over a mile, and Crystal Lake (24-acres), a little over a mile and one-half away.[3] After lunch and a nap, I hiked around the lake to the north, and saw a Mule deer, Yellow-belly marmots, a Yellow-rumped Warbler, and several White-crowned Sparrows.

The north end of Lawn Lake has the inlet that flows in from Crystal Lake, and a fairly large sandy delta where the creek comes in. This area

*North inlet and high-water line seen with white rocks.*

would have been underwater when the reservoir was 48-acres in size. At the far shore in the photograph below, the white rocks represent the high-water line of the reservoir, willows have filled much of this area in below that rocky line. In the next photograph from the north end of the lake, it is easy to visualize the larger basin of the old reservoir. Walking around the shoreline, the high-water line of the former reservoir is very distinct.

As I documented the lakeshore, I visited with a group of campers who were getting ready to hike out. They had one night left on their backcountry permit, but had been soaked with rain the night before, and didn't want a repeat of the weather for a second night. The Lawn Lake Campsites are on the east side of the lake and well above it, as they were in 1982, when the dam burst. The campers at Lawn Lake in 1982 had heard noises described as like wind rustling through the trees in the early morning (with no wind), which was likely noise from the dam failing (See Chapter 4). I finished my photography of the lake and headed down trail myself.

*The eastern shoreline, well below the dammed reservoir height.*

*View from the north of the Lawn Lake Basin.*

*My favorite Lawn Lake photograph.*

The hike out was slow, the group I followed wanted to stop at every flood feature along the way. One of the campers was writing a blog about the hike, including global positioning (GPS) locations for every junction and feature along the way.

When I got to the trail-head, I hiked over to the alluvial fan along the road. The trail was closed due to construction work at the site.

*top: View to the north from the trail. right: The trail held up by tree roots.*

What was noteworthy was the number of new trees growing in the now 32-year-old scar, and the huge boulders that had washed down the hillside. I would revisit this alluvial fan once the area had been restored as an interpretive site.

At the end of the day, I was tired, but the hike was important as closure for seeing this group of lakes in Rocky Mountain National Park without their historical dams.

## The Lawn Lake Alluvial Fan

The Lawn Lake Flood created an alluvial fan when the dam broke and the Roaring River tore down the mountainside. As it came down the west side of Bighorn Mountain and over Horseshoe Falls, it dumped tons of rocks, trees and soil down the hillside into Horseshoe Park as it joined Fall River. One rock was estimated at 452 tons.[4]

*Alluvial Fan, Horseshoe Falls*

*Lawn Lake Alluvial Fan, July 10, 2013*

The alluvial fan was 42 acres in size and as deep as 44 feet. It also created a new lake, named Fan Lake, with an area of 17 acres. In 1995, high water flows started to erode the dam that created the lake, and the National Park Service breached the lower part of this dam to drain the lake. This was done to lessen the risk of another dam failure.[5] A closer view of the alluvial fan and Horseshoe Falls is seen below. Both views were taken before the September 2013 floods, which further eroded the damaged Roaring River drainage.

*Lawn Lake Alluvial Fan, Horseshoe Falls, July 10, 2013.*

In the time frame between the 2013 floods and the summer of 2020, Rocky Mountain National Park rebuilt the Endovalley Road, and created parking lots and interpretive trails to allow visitors to visit this unique natural wonder, which was actually created due to the failure of a human-built structure, the Lawn Lake Dam.

We went to visit the newly reopened Lawn Lake Alluvial Fan area on September 30, 2020. RMNP had just reopened it with a new trail paved with FilterPave™, which is solid on top, but porous to provide drainage. The hike up this trail is steep for just the first part, a grade of 17.5 %.

This is an exciting front-country interpretive area, good parking areas, and an excellent chance for visitors to see and learn about this event.

*FilterPave™ trail lined with rocks.*

*A steel base bridge with more natural wooden deck and rails.*

*Roaring River-Horseshoe Falls coming down the Alluvial Fan.*

The trails offer great access to see the large rocks and boulders that came down the hillside, and the trails incorporate the flood strewn rocks in the design of this feature. The views are fantastic, either looking at the falls or toward Horseshoe Park. The new trees growing in the alluvial fan since 1982 add to the beauty, including the Aspen with fall colors.

*Abert's Squirrel*

An Abert's squirrel (Tassel-eared) scampered across the parking lot as I tried to photograph it, chased by a car, which ruined a chance for a better shot. Abert's squirrels, named for naturalist John James Abert, have nine subspecies and many variations in color. They feed on Ponderosa pine nuts, but do not cache seeds like other squirrels, so they must forage year-round. These were frequent visitors to my yard at the Dave Stirling Cottage in 1976-1977.

The Lawn Lake Alluvial Fan Interpretive Area is a great feature to visit. It is located by turning on the Endovalley Road off Fall River Road (Highway 34) and going past the Lawn Lake Trailhead toward the Endovalley Picnic Area and Old Fall River Road.

*Horseshoe Falls, relative scale of boulders to visitors.*

# Conclusion

Census data when water diversions began in Rocky Mountain National Park in 1890, listed the Colorado population at 413,248 people. The need for water was primarily irrigated agriculture on the eastern plains. When Rocky Mountain National Park was created in 1915, most of the dams and diversion projects in and around the Park were completed, except in the case of the Grand River Ditch, still under construction until 1937. This resulted in assurances when the Park was set aside, that these projects would remain part of the landscape. However, the Organic Act of 1916 establishing the National Park Service, designated protections within the Park that limited activities that would be detrimental to the Park's Mission. The Organic Act in part reads: *"The service thus established shall promote and regulate the use of the Federal areas known as national parks, monuments, and reservations…. which purpose is to conserve the scenery and the natural and historic objects and the wild life therein and to provide for the enjoyment of the same in such manner and by such means as will leave them unimpaired for the enjoyment of future generations"*.[1]

This mission statement directed National Park superintendents to manage their Parks based loosely on the "Protect, Use, and Enjoy" motto.

As people flocked to Rocky Mountain National Park, it was "suffering from a lag in funds and manpower for general maintenance and modernization of physical facilities".[2] This trend continued for decades, as the Park Service was inundated with visitors, and few funds were allocated over the years. Mission 66 was a new program implemented by the NPS in 1956, to fix and refurbish all the National Parks. Mostly used to enhance facilities for the automobile-oriented visitor, this was a great boon to enhance RMNP facilities and roads. Purchasing privately held land in the Park was part of this mission and resulted in many resorts and inholdings being removed from RMNP.[2] However, this did not take care of privately held water rights, or impoundments and diversions that deteriorated over time in the Park with owner neglect.

Once water projects were done, water managers typically did the minimum to maintain their projects, since water was delivered to farmers as cheaply as possible, and maintenance in the Park was limited, as far as access. I saw that when I worked for the Water Supply and Storage Company. In my opinion, this led to deterioration of storage and delivery systems in the Park. The Park Service had the desire to do something with these projects, but typically didn't have the funds to go forward.

*Brook trout in Boulder Brook Creek, Rocky Mountain National Park.*

An example of that was my boss talking to me about removing the Pear Lake dam in 1974. It was a pipe dream, but not a reality at the time.

My trails bosses, foremen, and our rangers were dedicated and professional National Park Service employees, with integrity in all their actions. That a lawyer implied that trail crew blasting activity was to blame, was the most ridiculous thing I had ever heard when I went to testify. I suspect the lawyers were seeking out those with the deepest pockets, by trying to blame the Federal Government. The National Park Service inherited the existing dams in the Park, where the intent of the law was to allow the owners and operators to carry out their duties, and also maintain dams and ditches under their chartered purpose.

As was written in the *Dam! Flood! Lawn Lake Dam Breach and Fall River Flood* book (1982): *"There is no malice here. Most probably not. Not in the men involved. Oversight and negligence, perhaps. A court will decide. Perhaps many courts over the years"*.

The courts did decide. The Park Service was only found partially responsible for the death of Terry Coates in Aspenglen Campground. All other liability for the Lawn Lake Flood rests solely on the former Farmers Irrigation Ditch and Reservoir Company. Improper caulking, and not encasing the valve housing in concrete 79 years before, during construction of the dam, unfortunately didn't aid the company in maintaining a properly working valve.

The National Park Service did the right thing afterward, by purchasing water rights and removing or rebuilding the remaining dams. By doing so, they have enhanced the safety of the visitors to the Park and those that live and work downstream. They have also restored the former reservoirs back to their near pristine nature, so that they may be enjoyed by visitors expecting a near wilderness experience. Water in Rocky Mountain National Park is the key to a quality park experience, whether it is a hiking to a stream, waterfall, high-country lake, or driving past the scenery created by this alpine mountain landscape.

Rocky Mountain National Park's and Colorado's precious water is under a great deal of pressure as the state outgrows its available resources.

Census data in 1950 listed Colorado's population at 1,325,089, with 415,795 people residing in the urban Denver, and 67,000 people in rural Weld County, which was primarily agriculture. Colorado census data is shown in Figure 1.

As the Colorado population grew over the decades, the majority of the population lived on the Front Range. As a result, cities and towns expanded their water needs for people rather than agriculture. Irrigated agriculture uses great amounts of water, especially for crops such as corn. High yield corn uses 600,000 gallons of water per acre, to produce 210 bushels of corn, which includes both rain and irrigation water.[3]

The City of Thornton purchased water rights from WSSC by "drying up" northern Colorado farms beginning in 1985. Their plans are to pipe this purchased water 74 miles, from the mouth of the Poudre Canyon, across Larimer and Weld Counties, and south to Thornton for the anticipated needs of their population in 2050.

| Colorado Census | Population |
| --- | --- |
| 1890 | 413,249 |
| 1950 | 1,325,089 |
| 1960 | 1,753,947 |
| 1970 | 2,207,259 |
| 1980 | 2,889,964 |
| 1990 | 3,294,394 |
| 2000 | 4,301,261 |
| 2010 | 5,029,196 |
| 2020 | 5,773,714 |

*Figure 1. Colorado Census Numbers from 1890 to 2020. U.S. Census Bureau.*

This is mostly Colorado River water that originates on the west side of RMNP. These senior water shares were purchased from the Water Storage and Supply Company. Thornton owns 287 shares of WSSC water, which has an estimated value of $400,000 a share. The market value of this water was between $86 million to $100 million dollars in 2005.[4]

The City of Greeley has initiated a project to store water for future needs, with plans to inject water into an aquifer under the Terry Ranch in northern Colorado. Opposition to this project has put measures on the ballot in 2021 to stop this project, citing potential contamination by Uranium in the area. The Windy Gap Project (see Chapter Two), through the Chimney Hollow Reservoir being built near Berthoud, should also provide Greeley for increased water needs. Chimney Hollow Reservoir will store up to 90,000 acre-feet of water.

Another water project to store for future needs in Northern Colorado, is the Northern Integrated Storage Project, or NISP. NISP plans to build two reservoirs and a series of pipelines to provide water to fifteen towns and water districts. It will provide up to 13 billion gallons of water annually, enough for 80,000 families on the Front Range. It will also build a small reservoir off the Cache La Poudre River near Highway 287, north of Fort Collins. Ditch water from the Poudre will fill that reservoir, which will then be pumped into Glade Reservoir. Glade Reservoir will be built where Highway 287 is currently located, which will require relocating 8 miles of highway, to the east. Glade Reservoir will hold up to 170,000 acre-feet of water. A second reservoir east of Eaton will be built, Galeton Reservoir, holding up to 45,600 acre-feet of water. This water will be diverted from the South Platte River, near the confluence of the Poudre River near Greeley. A series of pipelines will move water around the project to supply water needs year-round throughout the project. The project is also supposed to provide minimum stream flows for the Poudre River of 18 cfs in winter, and 25 cfs in summer for fish and wildlife, near and adjacent to the Fort Collins area.[5]

The use of Hydraulic Fracturing (fracking) techniques for extracting oil and gas have also become an increasing water need, with the majority of Colorado oil and gas production found in mostly agricultural Weld County. Hydraulic fracking uses between 1.5 million to 16 million gallons of water per well, much of which is not directly recovered.[6,7] Although fossil fuel use should be headed for a decline due to climate change, this is not likely in the near future for Colorado's Front Range oil and gas industry.

In 1994, Federal reserved water rights were decreed for all unappropriated water in Rocky Mountain National Park draining east of the Continental Divide. In 2000, absolute water rights for the Colorado

River drainage were given to RMNP, giving them water rights dating back to 1915.[8] However, owners of pre-1915 water rights still receive priority as senior holders, under the Prior Appropriations water rules. In recent years, minimum stream flows for fish and wildlife have been obtained for many watersheds, but in drought and low flow years, still remains a lower priority throughout the west.

Reservoirs can add storage to the water delivery system for the Front Range, but increasing periods of drought can cut allocations, and senior water users have priority under the Right of Prior Appropriation. The Colorado-Big Thompson project never distributes a full allocation in any given year, their usual allotment is 70%, but in a dry year that quota may drop to 50%. That quota may change during the water year.

Severe drought cycles are increasing in the west, which combined with catastrophic wildfires, can heavily influence water resources. Wildfires in 2020, adjacent to and burning in RMNP, the Cameron Peak and East Troublesome Fires, caused mudslides and particulate contamination to water supplies. For 39 days in 2021, the City of Greeley couldn't use normal water supplies due to burn scar runoff into their storage system.

In the Colorado River Basin, there are historically low reservoir levels in Lake Mead, and Lake Powell due to drought. As levels continue to drop, Lower Colorado Basin states Nevada, Arizona, and California may get shorted on their 7.5 million acre-feet allotment. If this drought period worsens, we are likely to see a re-adjudication of the Colorado River Compact.

The City of Greeley Colorado uses 0.5 acre-feet per family per year as their estimate of how much water use to plan for. The Town of Wellington Colorado uses 0.58 acre-feet per family per year as their estimate for planning new subdivisions. When I calculated our household water use for the last year, we used 0.31 acre-feet or 102,400 gallons. Although we have a native grass lawn, if that figure is used for our 206-house subdivision, that water use is 21.1 million gallons per year for less than 160 acres. Using the Greeley per household estimate, that would be 33.6 million gallons of water, and for Wellington estimates, 39.9 million gallons of water per year potentially used by our subdivision.

In the end, water is the life blood for the ecosystems of the National Park, and those that live and work downstream. These precious resources need to be protected, used efficiently, and carefully allocated.

Whether we can resolve the increasing human impacts on the Park from area population increases, high visitation, and automobiles smothering the Park, and ultimately, climate disruption and climate change, these will be the challenges of the present and the future.

For Rocky Mountain National Park, the mission statement of the Organic Act comes back to mind. It is worth repeating here:

> *"The service thus established shall promote and regulate the use of the Federal areas known as national parks, monuments, and reservations.... which purpose is to conserve the scenery and the natural and historic objects and the wild life therein and to provide for the enjoyment of the same in such manner and by such means as will leave them unimpaired for the enjoyment of future generations".*[1]

# Glossary and Acronyms

**Acre-foot:** The amount of water one foot deep (cubic foot) that covers an acre of land, which is 43,560 cubic feet or 326,000 gallons of water, which also weighs 2.6 million pounds. This is the amount of water often described that a typical family of four uses in a year.

**Alluvial Fan:** Delta-shaped deposits of sand, clay, silt, gravel, rocks and debris deposited at the mouth of a river, or the terminus of a flood location.

**Army Corps of Engineers (ACE):** The engineering branch of the military, which includes civilians and military personnel, providing expertise especially pertaining to waterways, dams, levees, and flood protection.

**Association of State Dam Safety Officials (ASDSO):** Group formed to study, discuss, and prevent future dam failures.

**Big Thompson (BT):** The Big Thompson River that flows out of RMNP, through Estes Park and then down the BT Canyon to Loveland. U.S. Highway 34 is the road that travels through the canyon.

**Bureau of Reclamation (BOR):** United States agency under the Department of Interior which builds, controls and manages dams and waterways.

**Civilian Conservation Corps (CCC):** Work force nationwide and in RMNP providing men with jobs related to public works.

**Colorado Department of Transportation (CDOT):** State agency responsible for maintaining and building roads.

**Colorado-Big Thompson Project (C-BT, CBT):** The large water project in Colorado built by the BOR, storing and diverting western slope water to the eastern slope for irrigation and power.

**Colorado State Patrol (CSP):** The state highway law enforcement agency.

**Creek:** Often a western term for a stream, or a brook. A creek is smaller than a river and usually empties into a river.

**Cubic Feet per Second (CFS, cfs):** A measure of waterflow within a stream or river, often reported during floods. It is equal to a volume of water one foot wide and one foot deep, flowing a distance of one foot in one second. One cfs is equal to 7.48 gallons of water flowing past a given location every second.

**Debris Flow:** Another term for landslide or mudslide. Also call a mass movement. A moving mass of loose mud, sand, soil, rock, and water that flows down a slope influenced by gravity.

**Doppler RADAR:** A Radio Detection and Ranging system using the Doppler effect to determine the location and velocity of a storm, clouds, precipitation, etc.

**Drainage:** Another term for watershed, that area of land that water drains into a stream or river.

**Farmers Irrigation Ditch and Reservoir Company (FIDRC):** Owner of the Lawn Lake Dam, and water rights from Lawn Lake.

**Federal Emergency Management Agency (FEMA):** The federal agency that provides disaster relief, such as following weather catastrophes, floods, etc.

**Federal Energy Regulatory Commission (FERC):** The federal agency that regulates energy, especially power generation and transmission.

**Five-Hundred Year Flood:** The definition of the statistical chance of a flood of a certain magnitude occurring once every 500 years.

**Flood:** Any abnormal fast, or explosive flow of water caused by an oversaturation of a creek or river, dam break, or an amount of water that leaves the normal creek bed or riverbank. Floods often form a wall of water and carry debris in their path.

**Hydrologist:** Someone who studies water, water flow, bodies of water, and soils associated with waterways.

**Mass Movement:** A mass movement is a landslide or mudslide created when saturated ground is unstable and mud, rocks, and trees move down a hill or mountainside. An alluvial fan is a type of mass movement in a stream bed or creek bed that is created during a flood. Mass movement may also be termed debris flow.

**National Interagency Fire Center (NIFC):** formerly BIFC, Boise Interagency Fire Center.

**National Oceanic and Atmospheric Administration (NOAA):** Department of Commerce agency concerned with monitoring oceans, the earth's environment, forecasting weather, hurricanes, and floods.

**National Park Service (NPS):** Agency that oversees National Parks, Historical Sites, National Recreation Areas.

**National Weather Service (NWS):** The part of NOAA that predicts weather.

**National Environmental Protection Act (NEPA):** This act requires federal agencies to conduct Environmental Impact Studies (EIS), or Environmental Assessments (EA) on projects on federal lands, prior to implementation of the projects.

**Northern Integrated Supply Project (NISP):** A project to build two reservoirs and a series of pipelines to provide water to fifteen towns and water districts in northeastern Colorado.

**One-Hundred Year Flood:** The definition of the statistical chance of a flood of a certain magnitude occurring once every 100 years. A one percent chance of a flood occurrence.

**One-Thousand Year Flood:** The definition of the statistical chance of a flood of a certain magnitude occurring once every 1000 years.

**Penstock:** An enclosed pipe, sluice, or floodgate that delivers water to a hydraulic turbine or sewerage system. Also may be a channel for conveying water.

**Reorganized Farmers Ditch Company (RFDC):** Formerly FIDRC. The owner of the water rights contained in Lawn Lake, reorganized after bankruptcy following the Lawn Lake dam failure.

**River:** A river is a flow of water larger than a creek or a stream that connects to a larger river and that then flows into a lake or an ocean.

**Rocky Mountain National Park (RMNP):** Established in 1915.

**State Engineers Office (SEO):** Colorado office that inspects, oversees dams in the state. State of Colorado Division of Water Resources (CDNR), Dam Safety Branch, State Engineers Office (SEO).

**Stream:** Another term for a creek, or a brook. Streams usually empty into a larger river.

**Tributaries:** A tributary is a river or stream or creek that flows into a larger river or lake, with increasing flows of water volume as these merge into larger rivers.

**United State Department of Interior (USDI):** Parent agency of the National Park Service.

**United States Geological Survey (USGS):** Science organization of the Interior Department.

**WAPA:** Western Area Power Administration.

**Water Rights: Law of Prior Appropriation:** Western U.S. water law.

*First Principle:* The public owns all the water
of the natural streams within the State of
Colorado. This prevents monopoly control of water supplies.

*Second Principle:* The water is dedicated to the use of the people through appropriation of unappropriated waters. Only the right of use can be obtained, and equal opportunity was the guiding principle.

*Third Principle:* Actual beneficial use of the public's water resource is required for the establishment of a water right.

*Fourth Principle:* When there is not sufficient water to satisfy all use rights, the earlier established right prevails.

*Fifth Principle:* All persons and corporations, upon payment of just compensation, are entitled to a right of way across private lands of another for the construction and operation of water works for use rights.

**Water Rights: Riparian:** Eastern U.S. water law. Riparian water rights are the rights that landowners have to make "reasonable use" of the water that abuts or flows through or over their properties. Riparian rights include the right to build structures like docks or piers, access to the water for the purposes of swimming or fishing, and the right to exclusive use of the water on their property if the water is not navigable.

**Watershed:** That area of land that water drains into a stream or river. A collection of the land and its streams that form a river.

# Endnotes

## Introduction

[1] Arps, L.W. and E.E. Kingery. 1977. *High Country Names. Rocky Mountain National Park,* Rocky Mountain Nature Association, Estes Park, CO. 212 pages.

[2] Reisner, Mark. 1993. *Cadillac Desert. The American West and Its Disappearing Water.* Penguin Books. 582 pages.

[3] United Nations IPCC. 2020. *Sixth Assessment Report. The Intergovernmental Panel on Climate Change.* August 21, 2021.

[4] Mann, Michael E. 2021. *The New Climate War: The Fight to Take Back Our Planet.* PublicAffairs, New York. 351pages.

## Chapter One: Water Resources

[1] *https://www.nps.gov/romo* Park information 2020.

[2] Paul, S.J. 2013. *Hiking Waterfalls in Colorado: A Guide to the States Best Waterfall Hikes.* Falcon Guides, Guilford CT.

[3] Dannen, K. and D. Dannen. 1982. *Rocky Mountain National Park Hiking Trails, Including Indian Peaks.* Third Revised Edition.

[4] USDI, USGS. 1961. Rocky Mountain National Park, Colorado, Topographical Map, Contour Edition. USGS, Denver, CO.

[5] Fogelberg, B. and S. Grinstead. 2006. *Walking into Colorado's Past, 50 Front Range Hikes.* Westcliffe Publisher, Englewood, CO.

## Chapter Two: Water Impoundments and Diversions

[1] Sprague, M. 1976. *Colorado: a bicentennial history. American Association for State and Local History.* WW. Norton and Co., Inc. New York, New York.

[2] Doesken, N.J., R.A Pielke, Sr., Bliss O.A.P. 2003. Climate of Colorado. Climatography of the United States No. 60. Colorado State University. *https://climate. colostate.edu/climate_long.html*

[3] Hobbs, G.J. Jr. and M. Welsh. 2020. *Confluence, the Story of Greeley Water.* City of Greeley, Colorado. 337 pages.

[4] Kissler, B.J. 1952. *A History of The Water Supply and Storage Company,* Colorado State College of Education, Greeley, CO.

[5] Buchholtz, C.W. 1983. *Rocky Mountain National Park, A history*. University Press of Colorado.

[6] Evans, H.E and M.A. Evans. 1991. *Cache La Poudre: the natural history of a Rocky Mountain River*.

[7] Fry, N.W. 1954. *Cache La Poudre, "The River", as seen from 1889 to 1954*.

[8] Watrous, Ansel. 1911. *History of Larimer County, Colorado*. The Courier Printing and Publishing Company, Fort Collins, Co. 297 pages.

[9] Hansen, J.E. 1991. *A Brief History of the Water Supply and Storage Company*. Colorado State University Water Archives, Fort Collins, CO.

[10] Photo Credits: Dave and Kathy Scranton. Dave 'n' Kathy's Vagabond Blog. 2017. Dave'n'Kathy's Vagabond Blog Rocky Mountain National Park-Colorado River Trail.html.

[11] Atkins, F. 1976. USDI, National Park Service, National Register of Historical Places Inventory-Nomination Form for Federal Properties.

[12] Rocky Mountain National Park. Fact Sheet: March 2010. Grand Ditch Breach Restoration-Environmental Impact Statement.

[13] Fogelberg, B. and S. Grinstead. 2006. *Walking into Colorado's Past, 50 Front Range Hikes*. Westcliffe Publishers, Englewood, CO.

[14] Autobee, Robert. 1996. *Colorado-Big Thompson Project*. Bureau of Reclamation. 37 pages.

[15] NCWCD. 2012. Colorado: Estes Powerplant. Discover our Shared Heritage Travel Itinerary Series. Northern Colorado Water Conservation District.

[16] Windy Gap Project: *https://www.northernwater.org/what-we-do/deliver-water/windy-gap-project*.

**Chapter 3: Dam Structures in Rocky Mountain National Park**

[1] National Register of Historic Places, National Park Service. 1984. E.O.11593. Determination of
Eligibility Notification. Bluebird Dam. 15 pages.

[2] McKnight, Patrick. *The Water Rights of Rocky Mountain National Park, A History*. Unpublished manuscript.

[3] Arps, L.W and E.E. Kingery. 1977. *High Country Names. Rocky Mountain National Park*, Rocky Mountain Nature Association, Estes Park, CO.

[4] Baker, Mark E. and McCormick, Bill. 2012. *30th Anniversary of the Lawn Lake Dam Failure: A Look Back at the State and Federal Responses.* National Park Service and Colorado Division of Water Resources, CO.

[5] Jessen, Kenneth. June 25, 2015. Many developed water storage in Rocky Mountain National Park. *Loveland Reporter Herald* newspaper.

[6] USDI, NPS, National Register of Historical Places. 2006. Lost Lake Trail.

[7] Lost Lake Water Rights. *https://irma.nps.gov/DataStore/DownloadFile/437325*

[8] Buchholtz, C.W. 1983. *Rocky Mountain National Park, A history.* University Press of Colorado.

[9] Bird, Isabella L. 1960. *A Lady's Life in the Rocky Mountains.* University of Oklahoma Press, Norman, OK.

[10] Nelson, R.A. 1970. *Plants of Rocky Mountain National Park.* Rocky Mountain Nature Association, Estes Park, CO.

[11] Mills, E. A. 1913. *Transportation Facilities, In Beaver World.* Houghton Mifflin Co. The Riverside Press, Cambridge, MA.

[12] National Park Service. 2011-12. Public Scoping Lily Lake Dam. *https://www.nps.gov/romo/learn/management/upload/public_scoping_lilylake_dam.pdf.*

**Chapter Four: The Big Thompson Flood of 1976**
[1] Wamsley, Sharlynn J. 2001. *Reflections on the River. The Big Thompson Canyon Flood.* Drake Club Press. 456 pp.

[2] Dollman, Darla Sue. 2017. *Colorado's Deadliest Floods.* History Press, Charleston, SC. 172 pp.

[3] Cotton, Don., Editor. 1976. *The Big Thompson Flood.* C.F. Boone, Publisher, Lubbock, TX. 63 pp.

**Chapter Five: Prelude to a Flood, The Lawn Lake Flood**
[1] Schmidt, H. 1982. Dailyaide calendar notebook. Henry Schmidt, Ditch Rider for the Farmers Irrigating Ditch and Reservoir Company.

[2] Baker, Mark E. and McCormick, Bill. 2012. *30th Anniversary of the Lawn Lake Dam Failure: A Look Back at the State and Federal Responses.* National Park Service and Colorado Division of Water Resources, CO.

[3] Cashman, S.R. July 15, 1982 Statement. Supplementary Case/Incident Record. Interviewer: Laurie Shannon, NPS.

[4] Brault, Margaret. July 15, 1982 Statement. Supplementary Case/Incident Record. Interviewer: Laurie Shannon, NPS.

[5] Wilde, Steven. July 15, 1982 Statement. Supplementary Case/Incident Record. Interviewer: Laurie Shannon, NPS.

[6] Wood, Tim. July 15, 1982 Statement. Supplementary Case/Incident Record. Interviewer: Laurie Shannon, NPS.

[7] Gillette, Steven W. July 22, 1982 Statement: Lawn Lake Flood. Supplementary Case/Incident Record.

[8] Davis, L.V, and Turner, Iris. July 20, 1982 Statement: Cascade Dam. Supplementary Case/Incident Record. Interviewer: Lisa C. Nichols, NPS.

[9] Thomas, Ingrid. July 17, 1982 Statement: Aspenglen Campground. Supplementary Case/Incident Record. NPS.

[10] Coates, Rosemary. July 17, 1982 Statement: Aspenglen Campground. Supplementary Case/Incident Record. NPS.

[11] Logan, Charles E., Ranger. July 19, 1982. Lawn Lake Dam Failure, subsequent search for missing persons. Supplementary Case/Incident Record, Case Number 820603. Pages 1-4. NPS.

[12] Box, David Allen. July 15, 1982 Statement: Lawn Lake. Supplementary Case/Incident Record. NPS.

[13] Hanson, Donay, Biological Aid, NPS. July12-14, 1982 Statement: Lawn Lake observations.

[14] Harpster, J.D. July 15, 1982 Memorandum, Subject: Lawn Lake Dam Failure in Rocky Mountain, NPS.

[15] Danielson. J. 1983. *Investigation of the Failure of Lawn Lake Dam, Larimer County, Colorado.* Office of the State Engineer (SEO), State of Colorado.

**Chapter Six: After the Flood, Liability and the Blame Question**

[1] Danielson. J. 1983. *Investigation of the Failure of Lawn Lake Dam, Larimer County, Colorado.* Office of the State Engineer (SEO), State of Colorado.

[2] Miller, Gardner. 1982. *Dam! Flood! Lawn Lake Dam Breach and Fall River Flood, Estes Park, CO.* July 15, 1982. QM Productions. VIP Printing, Estes Park. October 1982.

[3] Colorado General Assembly. 1981. Digest of Bills enacted by the 53rd General Assembly of the State of Colorado. 1981-First Regular Session. Office of Legislative Legal Services.

[4] Baker, Mark E. and McCormick, Bill. 2012. *30th Anniversary of the Lawn Lake Dam Failure: A Look Back at the State and Federal Responses.* National Park Service and Colorado Division of Water Resources, CO.

[5] Jarrett, Robert D. and John E. Costa. 1983. *Hydrology, Geomorphology and Dam-break Modeling of the July 15, 1982, Lawn Lake Dam and Cascade Lake Dam Failures, Larimer County, Colorado.* USGS Report 84-612.

[6] *Estes Park Trail-Gazette.* October 2, 1982. Faulty Gate Valve is Suspected Catalyst in Lawn Lake Flood. Jackie Hutchins.

[7] *The Denver Post.* March 3, 1983. Estes Flood Blamed on Pipe Flaw. P1-A and P10-A. Neil Westergaard.

[8] *Estes Park Trail-Gazette.* July 23, 1982. 'Super-lawyer' Gerry Spence to Probe Post-flood Resources. Tim Asbury.

[9] *Rocky Mountain News.* March 23, 1983. Bill Would Hike Flood Victim Reimbursement.

[10] *Rocky Mountain News.* April 15, 1983. Flood-suit Bill is Withdrawn in Senate Panel.

[11] Asbury, Tim. July 19, 1991. Absence of Proof releases Park Service of Dam Liability. *Estes Park Trail-Gazette,* P. 3.

[12] *Estes Park Trail-Gazette.* July 4, 1984. Developer's file $5 Million Claim Linked to Lawn Lake Dam Break.

[13] *Estes Park Trail-Gazette.* March 11, 1987. Flood Victims must pay for Faulty Lawsuit.

[14] *Estes Park Trail-Gazette.* May 13, 1988. 1982 Flood Claims Remain Unresolved.

[15] *Estes Park Trail-Gazette.* July 19, 1991. Court Spikes hopes for Lawn Lake Flood Claims.

[16] U.S. District Court for the Central District of Illinois. May 6, 1985. Coates v. United States, 612 F. Supp. 592 (C.D Ill. 1985).

**Chapter Seven: Dam Removal or Repair?**

[1] Baker, Mark E. and McCormick, Bill. 2012. *30th Anniversary of the Lawn Lake Dam Failure: A Look Back at the State and Federal Responses.* National Park Service and Colorado Division of Water Resources, CO.

[2] *Estes Park Trail-Gazette.* July 23, 1982. Ditch Company Still Unsure about Dams Future Status. Dan Campbell.

[3] *Estes Park Trail-Gazette.* July 23, 1982. Article: Park Service Examines Dams. Jackie Hutchins.

[4] *Estes Park Trail-Gazette.* July 28, 1982. Article: Park Service Inspects Dams in RMNP. Jackie Hutchins.

[5] *Estes Park Trail-Gazette.* August 18, 1982. Article: Park Inventories Remaining Dams. Jackie Hutchins.

[6] *Estes Park Trail-Gazette.* July 23, 1982. Article: Ditch Company Still Unsure about Dams Future Status. Dan Campbell.

[7] *Fort Collins Coloradoan.* March 25, 1983. Dam Inspection Plan Revised. Page B4.

[8] Colorado General Assembly. June 30, 1983. 1983 Digest of Bills, Office of Legislative Legal Services.

[9] Lost Lake Water Rights. *https://irma.nps.gov/DataStore/DownloadFile/437325*

[10] National Parks Conservation Association. 2009. *Restoring High Country Lakes, Benefits Native Species and Public Safety, in Responding to Climate Change in National Park: Economic Benefits of Restoring Natural Systems.* Rocky Mountain West Fact Sheet.

[11] National Register of Historic Places, National Park Service. 1984. E.O.11593. Determination of Eligibility Notification. Bluebird Dam. 15 pages.

[12] Karpowicz et al. 2008. *Removal of Bluebird Dam at Rocky Mountain National Park Colorado,* pages 1171-1184 in Collaborative Management of Integrated Watersheds.

[13] Rocky Mountain National Park. September 15, 2017. Decision Reached on Sprague Lake Dam Repair. Kyle Patterson, Public Affairs Officer.

[14] *https://svlhwcd.org/water-resources/.* Saint Vrain Left Hand Water Conservation District. December 19, 2007. Agreement with the National Park Service for Management of Copeland Lake Property.

[15] Colorado 14ers. September 22, 2013. Trailhead Info and Status (Copeland Lake Trailhead). Note posted on website regarding closure.

[16] Davis, D.W. and Jack Odem, June 4, 1951. Investigation of Lily Lake Dike Break and Resulting Flood, agency unknown.

[17] National Park Service. 2011-12. Public Scoping Lily Lake Dam. *https://www.nps.gov/romo/learn/management/upload/public_scoping_lilylake_dam.pdf*

[18] Baker, Mark E. 2014. *Dodging a Bullet: Lily Lake Dam and the 2013 Colorado Flood.* Presentation at Association of State Dam Safety Officials 2014 National Conference.

[19] *Loveland Reporter-Herald.* November 20, 2018. Fish Creek Flood Repairs in Estes Park Named Top Disaster Recovery Project. Pamela Johnson.

[20] National Park Service. 2012. Information brochure on Lily Lake. *https://www.nps.gov/romo/lily_lake.htm*

[21] Town of Estes Park. March 9, 1994. Estes Park Municipal Hydroelectric Station, Declaration of Intention. Letter to the Secretary Federal Energy Regulatory Commission, including a Position Statement from the Rocky Mountain National Park Superintendent.

[22] USDI. NPS. April 16, 1998. Stanley Hotel District, Stanley Powerplant addition.

[23] Town of Estes Park. May 12, 2011. Cascade Diversion Water Rights Report.

[24] *Estes Park Trail-Gazette.* July 9, 2009. Historic Fall River Hydroplant open for interpretation.

[25] Rocky Mountain Conservancy. 2017. Cascade Cottages Campaign. 2015 Cascade Cottages Campaign —Rocky Mountain Conservancy (*rmconservancy.org*).

### Chapter Eight: The 2013 Colorado Front Range Floods

[1] Batka, John. 2014. Colorado Dam Safety Branch. *From Response to Recovery-The Story of the 2013 Colorado Floods from the Perspective of the Members of the Colorado Dam Safety Branch.* ASDSO Annual Conference. San Diego, California: Association of State Dam Safety Officials.

[2] Dollman, Darla Sue. 2017. *Colorado's Deadliest Floods.* History Press, Charleston, SC. 172 pp.

[3] Yochum, S.E. and D.S. Moore. December 2013. *Colorado Front Range Flood of 2013: Peak Flow Estimates at Selected Mountain Stream Locations.* US Department of Agriculture, Natural Resources Conservation Service.

[4] Baker, Mark E. and McCormick, Bill. 2012. *30th Anniversary of the Lawn Lake Dam Failure: A Look Back at the State and Federal Responses.* National Park Service and Colorado Division of Water Resources, CO.

[5] Heindel, Jennifer. 2014. *Front Range Floods Teacher Guide.* Rocky Mountain National Park. 23 pages.

### Chapter Nine: Return to the De-dammed Lakes

[1] Gossett, Daniel. July 29, 2016. After the Flood: New Ouzel Falls bridge practical, but misses point. Guest Columnist, Adventure Section, *The Greeley Tribune* newspaper, pages B1-B2.

[2] Mills, Enos. 1909. *Colorado Snow Observer, in Wildlife on the Rockies, republished in Enos Mills' Colorado.* Pickering, James H. 2006.

[3] Dannen, K. and D. Dannen. 1982. *Rocky Mountain National Park Hiking Trails, Including Indian Peaks.* Third Revised Edition.

[4] Dollman, Darla Sue. 2017. *Colorado's Deadliest Floods.* History Press, Charleston, SC. 172 pp.

[5] Baker, Mark E. and McCormick, Bill. 2012. *30th Anniversary of the Lawn Lake Dam Failure: A Look Back at the State and Federal Responses.* National Park Service and Colorado Division of Water Resources, CO.

**Conclusion:**

[1] USDI. 2005. NPS Organic Act. National Park Service Organic Act and its Implementation Through Daily Park Management. *https://www.doi.gov/ocl/nps-organic-act*

[2] Buchholtz, C.W. 1983. *Rocky Mountain National Park, A history.* University Press of Colorado. Page 202.

[3] Comis, Don. August 2011. Agricultural Research Service, USDA. Growing Crops and Saving Water in the West. *Agricultural Research magazine,* ARS, USDA.

[4] Miller, Chase. September 16, 2005. City of Thornton ready to get out of the land business – *BizWest magazine.*

[5] NISP. *https://www.northernwater.org/nisp*

[6] *USGS.gov.* Energy. Web Link: How much water does the typical hydraulically fractured well require? (*usgs.gov*).

[7] Carter, J.M., Macek-Rowland, K.M., Thamke, J.N., Delzer, G.C., 2016, Estimating national water use associated with unconventional oil and gas development: U.S. Geological Survey Fact Sheet 2016–3032, 6 p. *http://dx.doi.org/10.3133/fs20163032*

[8] RMNP, National Park Service. Hydrologic Activity Fact Sheet.

# Index

# Daniel N. Gossett

Dan began his love affair with Rocky Mountain National Park in the 1960's, hiking and camping in Wild Basin, where water was an important part of the experience. He started working for the National Park Service on seasonal trail crews and as a trail's foreman beginning in 1974, through 1980. He was on the RMNP Technical Rescue Team from 1976-1980 and fought the Ouzel Fire from its overnight "blowup" and then for over a month "mopping it up". As well as trail crew, he worked at the Park as a volunteer doing natural resource photography, Park dispatch, issuing backcountry permits, and manning the information desk at Park Headquarters.

Dan worked as a paramedic, and as a seasonal Colorado State Park Ranger in the 1980's.

As a wildlife student, his wildlife biology career started with sampling vegetation plots with Tom Hobbs on Tom's PhD project on elk in Rocky Mountain National Park in 1976. After completing an A.A.S. in Veterinary Technology, his B.S. in Biology at Colorado State University, and an M.S. degree in Raptor Biology at Boise State University, he worked with a variety of wildlife species.

Dan worked with bison, raptors, desert bighorn sheep, and neotropical migrant songbirds. On the Duck Valley Indian Reservation, he completed a survey of Greater Sage Grouse, and conducted a Blue Creek Wetlands study. After discovering West Nile virus in sage grouse there, he worked to determine its impact on the reservation's grouse population.

He concluded his federal career at the USDA National Wildlife Research Center in Fort Collins, Colorado, working with a large variety of wildlife, before retiring to pursue writing and photography.

# Acknowledgements

The following people have aided me creatively, technically, logistically, or emotionally in creating this book.

My life partner, Stephanie Lynn Gossett, whose encouragement, editorial skills, love, and support have made both my writing and life worthwhile.

My parents, B. Newton and Sue P. Gossett, who have supported all my efforts throughout the years and encouraged my writing, as well as my naturalistic endeavors and photography.

The employees of Rocky Mountain National Park in the 1970's, for helping burn in my love for the Park. Kelly Cahill assisted me with various NPS collections and materials for this effort. Kathy Brazelton, Kevin Soviak, and Joe Arnold assisted with a review of subject matter for this book. Joseph Chafey and Ken Czarnowski helped with water rights. Reporter Jackie Hutchins helped with newspaper articles.

To Bruce Haak, and Kate Davis for their advice on publishing, and Karen Steenhof for her teaching on scientific writing. To Mary Lu Gorsline for teaching Composition in the Eighth Grade. My apologies to E.B. White and William Strunk Jr., where I have violated the rule of being concise.

The Rocky Mountain Conservancy (RMC) promotes stewardship as a partner with Rocky Mountain National Park, through education, award winning interpretive publications, and philanthropy. For more information or to donate, go to *RMConservancy.org*.